D1223062

The One-Minute Workout

The One-Minute Workout

Science Shows a Way to Get Fit That's Smarter, Faster, Shorter

Martin Gibala, PhD

With Christopher Shulgan

AVERY
An imprint of Penguin Random House
New York

AVERY
An imprint of Penguin Random House LLC
375 Hudson Street
New York, New York 10014

Most Avery books are available at special quantity discounts for bulk purchase for sales promotions, premiums, fund-raising, and educational needs. Special books or book excerpts also can be created to fit specific needs. For details, write SpecialMarkets@penguinrandomhouse.com.

Library of Congress Cataloging-in-Publication Data

Names: Gibala, Martin, author. | Shulgan, Christopher, author.
Title: The one-minute workout : science shows a way to get fit that's smarter, faster, shorter / Martin Gibala ; with Christopher Shulgan.
Description: New York, New York : Avery, 2017.
Identifiers: LCCN 2016036523 | ISBN 9780399183669 (hardback)
Subjects: LCSH: Interval training. | Physical fitness. | Exercise. | BISAC: HEALTH & FITNESS / Exercise. | HEALTH & FITNESS / Health Care Issues. | SCIENCE / Life Sciences / Human Anatomy & Physiology.
Classification: LCC GV481 .G48 2017 | DDC 613.7—dc23
LC record available at https://lccn.loc.gov/2016036523

Printed in the United States of America
1 3 5 7 9 10 8 6 4 2

BOOK DESIGN BY TANYA MAIBORODA

The material in this book is for informational purposes only. As with any new or changed exercise regimen, the programs described in this book should be considered only after consulting with your physician to determine which might be appropriate for your individual circumstances and the current state of your health. The intensive nature of some of the protocols in this book may involve a higher degree of risk for certain individuals, and your physician should help you determine if that risk is appropriate for you. Although every effort has been made to provide the most up-to-date information, the medical science in this field and information about high-intensity interval training are still developing. The authors and publisher expressly disclaim responsibility for any adverse effects that may result from the use or application of the information contained in this book.

Contents

The One-Minute Workout

CHAPTER ONE

Fit in Just Minutes a Week?

FEEL LIKE YOU DON'T HAVE TIME TO EXERCISE? LOOKING for a way to get in shape—fast? Of course you are. Regular physical activity makes you look and feel better. You'll also fight the aging process, go through your days in happier spirits, and reduce your chance of developing ailments like cardiovascular disease, diabetes, and even cancer.

I think exercise is one of the best things around. Most of us are under the impression, however, that exercise has to be time-intensive. We have this notion that it takes at least an hour to get in a good workout—*more* if you factor in the time required to get to and from the gym.

My studies show that idea is nonsense. The past decade

has seen an explosion of research into the science of high-intensity interval training, better known by its acronym, HIIT, pronounced "hit." We're learning that HIIT can provide serious benefits that increase a workout's time-efficiency. Sprint interval training, or SIT, which is the most extreme version of the technique and is characterized by a few brief bursts of all-out exercise, is especially potent. We're not just talking running here. HIIT techniques can be applied to virtually any mode of traditional cardio-type exercises, such as cycling, swimming, or rowing. Thanks to the new science of ultralow-dose exercise, those who read this book will learn strategies to get fit in the time required to grab a coffee, update a Facebook status, or check a Twitter feed.

Think for a moment about the traditional concept of what it takes to get fit. Most of us will envision an activity that requires hours and hours of hard work. Lots of miles pedaling in the bike saddle. Entire afternoons navigating running trails. Lap after lap at the local pool. Consequently, many people are too intimidated to even *try* to get fit. Many of us feel like there simply isn't enough time to fit in a workout.

But you know what? That's *wrong*. That's what my years of study have taught me. I've discovered that fitness is possible without spending countless hours in the gym. I don't want to say that all the people who do that are wasting their time. But the fact is, a method exists that enables you to reap the benefits of *hours* of exercise in just *minutes* per day. Strategies can be incorporated to transform you from out of shape to fit in the least amount of time possible. Among my biggest discoveries is a workout that provides the benefits of nearly an hour of steady aerobic exercise with just a single *minute* of hard exercising.

Pretty remarkable, right?

This book is for the people who believe they don't have time to exercise. In these pages, I describe techniques pulled from the latest scientific studies—how they work and how you can use them. I also provide tips on how to best manage your weight. And provide some easy methods to design muscle-building workouts that can be conducted anywhere from hotel rooms to your local park, with little or no need for special equipment.

I know, I know—many personal trainers and workout celebrities promise such benefits. But they don't have the deep scientific knowledge that comes from being a leading researcher in the field. The groundbreaking studies that have come out of my lab have been covered by the *New York Times*, *Time* magazine, and *NBC Nightly News*, to name just a few media outlets. In 2015, a review article that I wrote on the topic of time-efficient exercise was the most accessed paper in the *Journal of Physiology*, the world's most cited physiology journal. In fact, the top two titles on the *Journal*'s annual ranking of most-accessed papers were from my laboratory, and we had three in the top fifteen. That's quite a feat, considering all the amazing human physiology research that is conducted worldwide.

I am also fortunate to work with a lot of great people. As the chair of the Department of Kinesiology at McMaster University in Hamilton, Canada, I interact on a daily basis with people whose assembly of brainpower ranks among the best on the planet. "McMaster is one of the centers of the universe when it comes to exercise," observes Carl Foster, a physiologist at the University of Wisconsin–La Crosse and a past president of the American College of Sports Medicine. In fact, McMaster is a world-leading center of excellence in the study of how physical activity changes human physiology and health.

With this book, I've drawn on the expertise of past and present McMaster minds and some of the smartest exercise physiologists in the world to create the most definitive guide to time-efficient exercise. I hope you'll also find it's an entertaining read. Once you've finished it, you'll know enough to design your own time-efficient workouts. And you'll have grasped the techniques required to go from out of shape to a buff portrait of health in the least amount of time.

Already fit? If you're not using the techniques described here, chances are you're getting beat by someone who is. This book will provide you with techniques that can help you break through a training plateau and drop seconds or even minutes from your personal-best times. It'll also allow you more time to do other stuff, like work or hang out with loved ones, because you're not spending hours in the saddle, on the trails, or in the pool. And during those weeks or months when work or other duties make it difficult to exercise, this book will provide you with a series of techniques designed to *maintain* your fitness level in the minimum amount of time.

So let's get to discussing the most time-efficient workouts

possible, for everyone from couch potatoes looking to get in shape to athletes wanting to boost their race performance. No longer do you have to fit your day around your workout. Now you can fit working out around your day.

Introducing a More Time-Efficient Way to Work Out

So what is interval training? Basically, it's bursts of intense exercise separated by periods of recovery, which can involve complete rest or lower-intensity exercise. Understanding the concept is easier if you contrast it with regular endurance training. That's the sort of thing most people envision when they think about heading out for a run. Or head out for a swim. Or a ride. The point is that traditional exercise training involves traveling a certain distance at a relatively constant pace. The resultant graph of effort versus time looks roughly like this:

That line of constant effort might stretch out to forty-five minutes, an hour, ninety minutes, or even more. When you can afford the time, it's wonderful to get out and just run or ride with your mind at ease. That sort of training has a lot of therapeutic benefits. It reduces stress and can provide the opportunity to enjoy the outdoors. But my research has shown that it is anything but the most efficient way to train.

If time is our most valuable resource, and if we're attempting to get the most benefit from exercise in the least amount of time, then, as my research has shown, we're better off employing interval-training techniques. The graph of an interval-training workout looks more like this:

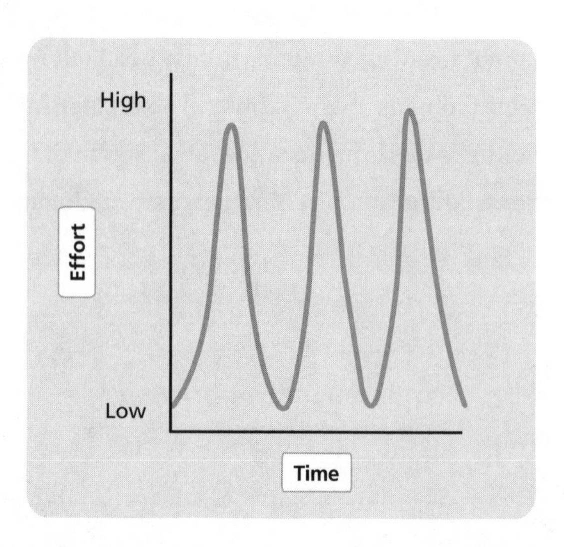

The idea is to vary the intensity of your workout. Go hard, relax, go hard, relax. The harder you go, the shorter the duration and the fewer intervals you need to achieve the same benefits of a much longer endurance-training workout.

People have been trying for centuries to get the benefits of exercise in creative ways that require less time. Think about hucksters wandering the Wild West promoting health elixirs or comic-book classified ads promising strongman muscles in mere weeks. More recently, the prestigious academic journal *Cell* published a study about a compound, known by the acronym AICAR, that helped sedentary mice run for 44 percent farther than untreated mice. The study raised a flurry of excitement about the possibility of developing an exercise pill, yet no one's been able to replicate the results in humans.

Interval training is the closest thing we have to an exercise pill. And over the past ten years there's been an explosion of research into the technique. This research has been conducted in my own lab as well as those of my colleagues all over the globe. And researchers like myself have concluded that high-intensity interval training may be the most efficient workout that the science of physiology has ever produced. Neatly summing up its benefits, A. J. Jacobs wrote in *Esquire*, "HIIT could be the biggest time-saver since microwaves."

HIIT is so popular that it has ranked at or near the top of the annual list of worldwide fitness trends compiled by the American College of Sports Medicine, the largest sports medicine and exercise science organization in the world. Personal trainers everywhere from New York to Hong Kong are staging fitness classes based on principles I helped establish at McMaster, just as Hollywood stars and Victoria's Secret models are using HIIT principles to get ripped for movie roles and fashion-week runway appearances. But here's the thing:

We've *still* got this idea in our heads that interval training is reserved for incredibly fit people working out in gyms in incredibly tight clothing. And doing workouts that last about an hour.

Again, that's nonsense. Workouts don't *have* to last an hour. They can last ten minutes or even less—and get you remarkable fitness benefits in that time. Even if you're overweight. Even if you're out of shape. There is a "flavor" of interval training appropriate for you. This exciting new science of interval exercise can be adopted to help everyone get fit—especially the people who long ago wrote off exercise because they felt like they didn't have the time.

Want to get fit fast? Or just get up a flight of stairs without losing your breath? Interval training can help. Want to cut your Ironman time? Burn fat faster? Or simply increase how far you can pedal on your Sunday ride? Interval training can help with that, too.

Most important, it can help in a lot less time than you ever thought possible. Interval training is perfect especially for the most time-pressed among us—everyone from city-hopping frequent-flyer executives to stay-at-home parents.

Public health guidelines generally call for at least two and a half hours per week of moderate-intensity exercise to gain health benefits. Devoted fitness enthusiasts often schedule at least an hour per workout. Interval training is a way to get the benefits of an hour-long run or bike ride in a fraction of the time. In the most extreme form, known as sprint interval training, it's possible to get those same benefits with just three minutes of hard exercise a week. In fact, my lab conducted the study that showed this—and the resultant news story rocketed to the top of the *New York Times'* most-popular-articles list.

I'm excited about this book for the same reason I'm excited about interval training itself—for their potential to make the benefits of exercise available to the greatest possible number of people. To that end I've tried to make this book a compelling instruction manual that's written in plain language and understandable for those without a degree in physiology.

I tell readers what interval training is, why and how it works, and who it works for. Then I provide a series of workouts and microworkouts that have been tested in laboratories around the world, followed by a discussion of the workouts' benefits, which have been established through rigorous scientific study.

The technique can be applied to pretty much any sort of exercise—and the most time-efficient versions include elements that boost both cardiovascular fitness and strength. Cycling, swimming, or bodyweight-style movements like burpees, push-ups, and pull-ups—they all accommodate interval-training techniques.

Now, I'm excited to translate the latest science into training techniques that virtually anyone can use.

How I Got HIIT

These days, when I conduct interviews on television shows or in newspapers, the journalists call me things like the "guru" of interval training. This makes me a bit uncomfortable, especially given the training method's incredible history (which we'll consider in chapter three). It's true that I've devoted most of my career to researching the topic. I've published dozens of academic studies in peer-reviewed journals over the past decade involving every aspect of interval training—how to do it, who can benefit, and how its potency compares with that of more traditional exercise approaches.

Looking back to the beginning of my research, I can see why I started pursuing it when I did. In 2004 I had just begun my all-important second three-year contract as an assistant professor at McMaster University. Tenure is a tough thing to secure in academia, and I had less than thirty-six months to prove myself as an asset to the university—or be jettisoned from the faculty. It was my career's ultimate make-or-break period.

On top of the pressure to produce quality research, I was teaching three courses, including one to more than two hundred undergraduates. My wife, Lisa, had just gone back to work as a high school physical education teacher. We had two young boys, ages one and three. Juggling my child-rearing responsibilities with teaching and research meant that, for the first time in my life, I felt like I didn't have time to exercise.

I can remember coming home from my office hours and walking through my front door excited at the possibility of trying to squeeze in a workout. But then something would come up. The boys would need to be fed. Or we'd be out of milk. Somebody might have a fever. I'd put off exercising to cater to more pressing needs. It would be days or even a week before another chance to exercise would emerge.

Interval training at that point was largely confined to the domain of highly trained individuals focusing on athletic performance. Few regular people ever performed interval-based workouts. Explaining why the average person didn't appreciate the value of intervals requires learning a bit about the way the body works.

Fitness means different things to different people. To exercise scientists, it means *cardiorespiratory* fitness, a parameter that can be measured in the laboratory by way of a test called maximal oxygen uptake or "VO_{2max}" (the "V" stands for "volume"). It is also called *aerobic* fitness, and it refers to the capacity of your body to transport and utilize oxygen. Scientists have found that it's one of the best predictors of overall health. The more aerobically fit you are, the better your heart can pump blood, the longer it takes you to get out of breath, and the farther and faster you're able to bike or run or swim. One more thing: It also happens to be the form of fitness that helps you live longer and live better by reducing your chances of developing ailments like cardiovascular disease and diabetes. Aerobic fitness is the thing most of us want when we first start working out.

So how do you build aerobic fitness? For a long time, many coaches and athletes thought that becoming aerobically fit required an enormous amount of exercise performed at a moderate pace. This thinking is reflected in the public health guidelines that generally call for 150 minutes of moderate-intensity exercise per week to derive health benefits. Two and a half hours. At minimum. While this represents less than 2 percent of the total hours in a week, it is a substantial commitment for most people, who cite "lack of time" as their main barrier to exercise.

The problem with these guidelines is that they scare plenty of people away from exercising. Lots of people are actually incredulous about it—it's like, you want me to work out for *two and a half hours a week*? Are you *crazy*? I can barely manage the laundry! And in fact, only 15 to 20 percent of Americans actually meet those fitness guidelines.

And yet, the message about how to get fit stayed largely the same: lots of steady-state exercise, which involves moderate activity performed at a constant rate for a prolonged period of time. In part, that's because people believed there just wasn't any other way.

According to the old thinking, sprints were thought to help your *sprinting*. They were a tool to boost *speed*—not aerobic fitness. Intervals might help an athlete develop a faster *sprint* time, but coaches and scientists figured they wouldn't help much in longer races. Nor would they help much with overall health or fitness. That, at least, was the conventional wisdom in the years before we arrived at our new understanding of the physiology of interval exercise.

A Fascination with Intervals

The old line of thinking has changed dramatically over the past ten years, thanks to an explosion of research into the benefits of ultralow-dose exercise. We now know that interval training absolutely can improve aerobic fitness and confer health benefits that we normally associate with substantially greater amounts of endurance training.

I came to this realization thanks in large part to one of the courses that I teach, a fourth-year elective called the Integrative Physiology of Human Performance. It focuses on the way the body's various systems—circulatory, respiratory, muscular—work together to meet the energy demand of exercise. Ever since I started teaching the course, my students have been interested in the training regimens of elite athletes. We would discuss everyone from the record-breaking four-minute-miler Roger Bannister to the Tour de France champion cyclist Lance Armstrong (this was before the doping scandal that eventually saw him stripped of his seven victories).

One training method common to both athletes was intervals—short, hard efforts performed repeatedly. My students found it difficult to understand why intervals would help such athletes as middle-distance runners and road cyclists. After all, the athletes' specialties very much depended on their *aerobic* fitness—on their ability to sustain physical activity without getting tired. The students would ask, "How can interval training boost your aerobic fitness when it is such an *anaerobic* type of exercise?"

The students' question was based on a misconception—one that many people have had for a lot of years. It has to do with the body's aerobic and anaerobic energy systems. I'll explain this in more detail later on, but for now, just remember that the body has two main ways of powering movement. It draws mostly on the *anaerobic* system when it requires lots of power, as when lifting heavy weights or all-out sprinting. And it mainly draws on the *aerobic* system when it performs less-intense movements for longer bouts of time, like jogging or cycling long distances.

But what about *repeated* sprints? It turns out that these are particularly taxing on the *aerobic* system. To illustrate this point to my students, I would show them the following graph, which reflected our evolving understanding of the energy required for repeated sprints. It's based on research conducted at McMaster in the late 1990s. In the study, subjects performed a series of three all-out thirty-second sprints on a bicycle, separated by four minutes of rest in between.

The left panel shows the energy distribution during the first go-as-hard-as-you-can sprint. Most of the energy is derived from the anaerobic system, although the contribution from the aerobic system increases over the course of the sprint. The right panel shows how that pattern changes over the course of the third sprint, with the aerobic system accounting

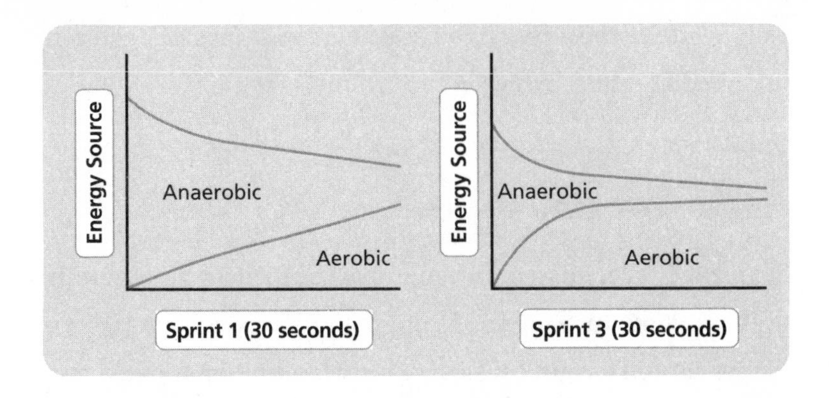

for a greater proportion of the energy. The precise energy distribution depends on many factors, including the duration of the sprints and the recovery periods, and how many work-rest cycles are completed. But the essential message is the same: as you repeat the basic pattern of sprint, rest, sprint, rest, a greater proportion of the energy comes from *aerobic* metabolism.

I started to wonder, what if people did *only* intervals? What benefit would that provide? How would it affect aerobic conditioning? Certainly, there were clues sprinkled throughout the scientific literature. A 1973 study of Swedish military recruits by one of my mentors, the legendary Scandinavian physiologist Bengt Saltin, concluded that physical fitness could be rapidly improved by interval training despite a short time investment. The Japanese researcher Izumi Tabata showed in 1996 that training using brief, intense intervals could substantially improve cardiorespiratory fitness. Two years later, my master's thesis adviser at McMaster, Duncan MacDougall, showed that training in short, hard intervals could dramatically increase the amount of mitochondria in muscles, the cellular bodies that use oxygen to burn fuels for energy.

A Life-Changing Study

So in 2004, with missed workouts leaving me feeling generally overwhelmed, time-pressed, and out of shape, I went into an intense brainstorming session with a handful of talented and enthusiastic graduate students. (Through the years, I've had the privilege of working with many such students who have contributed greatly to my research, and whose names are included on many of the papers I've published, along with those of my other research collaborators.) I wanted to study the effects of interval training and whether it was a more efficient way to exercise. Could intervals provide all the benefits of a long run or bike ride—in just a fraction of the time? During the brainstorming session, we decided to focus on how *little* exercise might be able to provide fitness benefits.

I was up at the chalkboard, writing furiously, while the students and I debated the designs of potential experiments. Envision lots of shouting while the chalkboard grew increasingly dense with graphs and equations, clouds of chalk dust swirling in the air, and you'll get close to the atmosphere in the room. How many intervals should there be? How long should each one last? How many rest days should come between them?

We came up with a simple, short, and focused experiment

to examine the power of high-intensity intervals. Thanks to the work of our fellow academics, we knew that intervals could boost cardiovascular fitness and the number of energy-producing mitochondria in muscle tissue. But we didn't know how quickly these adaptations took place, nor did we know how little exercise was required to trigger the benefits. We also didn't have a good sense of the power that sprints had to improve performance during good, old-fashioned steady-state aerobic exercise.

The experiment we designed looked at whether a handful of sprints could improve endurance performance. It worked like this: We assessed how long the subjects could pedal a stationary bicycle set to a fixed workload. Then the subjects went off and conducted six training sessions over two weeks.

The six training sessions conducted by the subjects required them to sprint on a stationary bicycle. The specific protocol is known as a Wingate test (named after the Israeli sports institute where it was developed, the Wingate Institute), which is designed to measure someone's *anaerobic* power—the all-out explosive effort necessary to bike, skate, run, or swim as fast as possible. You hop into the saddle and pedal as hard and as fast as you can for thirty seconds against a high resistance. The idea is to go all out. "Go as hard as you can," I told the subjects. "As if you're sprinting to save a child from an oncoming car—go that fast."

Wingates can be helpful. "If you've never done the Wingate-cycle test, let me try to explain what it feels like,"

A. J. Jacobs wrote in his *Esquire* article about interval training. "It feels like your legs are giving birth. It feels like you've got an eight-martini hangover in your calves. Your face contorts like a porn star in an AVN-award-winning threesome scene. You emit noises that resemble feedback at a thrash-metal concert. . . . The upside: It's over in 30 seconds." Jacobs is exaggerating just a little about how painful Wingates are. But his greater point is sound. You're supposed to give it your all.

Our experiment's eight subjects were the sort of athletic young adults, men and women, who tend to be around McMaster's Kinesiology Department. They participated in athletic activities in some form a couple of times a week but weren't involved in any kind of structured training program. The first workout of the study entailed four Wingates, or four rounds of cycling all out for thirty seconds each, with four minutes of rest in between. The training sessions took place in the lab at the Exercise Metabolism Research Group—a wide, low-ceilinged room filled with lots of computers, monitors, breathing tubes, and exercise equipment like stationary bicycles and treadmills. It wouldn't look out of place in *Blade Runner* or the science fiction movies of Neill Blomkamp.

The training sessions were pretty intense. In our write-up of the experiment, we noted that the subjects were "verbally encouraged" during their sprints. That's a staid depiction of what actually happened. The atmosphere was as loud and enthusiastic as any I've seen in a laboratory setting. Rock music blared. As each participant shifted into the sprint, a half-dozen grad students gathered around to offer encouragement: High-volume shouts. Lots of "Go! Go! GO!" and "YOU CAN DO IT!" Then, after the sprint, there were high-fives and pats on the back.

The most exciting day happened once the two weeks of training were over, when we conducted the final part of the experiment. Remember how, before training started, we asked our subjects to pedal a stationary bicycle set to a fixed workload for as long as they could? Well, now we asked them to do it again. This was the study's key measurement, where we assessed the potential performance benefit of the sprint-training sessions. I didn't have any idea how the subjects would perform or even whether there would be *any* benefit. We were testing the outcome of only six sessions of sprint training. The total time spent exercising in our study was only sixteen minutes. Would the training really do anything for endurance performance?

One by one, the eight sprint-training subjects conducted their final tests. The lab was quiet as they did. In the interests of objectivity, we didn't offer our riders any encouragement or feedback for this part. We also tried to keep our expressions blank, so as not to affect the subject's performance.

As the results came in, it was difficult to maintain that façade of impartiality. The numbers were crazy. The sprinters had *doubled* their endurance times. On average, before the training, the eight subjects could pedal the bike for twenty-six minutes until exhaustion. Following the six sessions of interval training, the average time was fifty-one minutes. It was an amazing result.

That was the moment I grasped how potent sprint intervals were—and how much they had the potential to improve overall fitness. It was incredible: in approximately the time required to do the dishes, these young men and women had *doubled* their endurance capacity. It was the most remarkable result I'd ever experienced in my lab.

The sprints changed the subjects' bodies in other ways, too. We obtained biopsies of their thigh muscles before and after training. The biopsies showed that, after the training, subjects had significantly more mitochondria in their muscles—because a key enzymatic marker, citrate synthase, increased by 38 percent. That matters because mitochondria are the muscles' powerhouses. We'll talk more about the physiology of training in chapter four, but in short, more mitochondria mean you can generate aerobic energy more quickly and with less fatigue.

It was the first time that such low volumes of training had been demonstrated to have such powerful effects. I marveled at the results on a number of different levels. There was something very powerful about going all out. About giving it your best possible effort. These short, intense intervals, a supposedly "anaerobic" exercise, appeared to have some sort of near-magic ability to improve aerobic energy metabolism. I couldn't get over how little exercise was needed to produce such enormous effects. A *doubling* of endurance capacity in only six training sessions? With just sixteen minutes of hard exercise?

It seemed miraculous. To this day, more than ten years later, I am still amazed by how little interval training you need to boost fitness and health.

And keep this in mind: We used a stationary cycle for our study for pragmatic reasons, but we figured the results would be the same for *any* activity conducive to a sprint approach, such as running, rowing, and even stair climbing. And in the ten-plus years that have passed since, this expectation has been borne out.

One of the top journals in the field of exercise physiology, the *Journal of Applied Physiology*, published the study on June 1, 2005. It also published a feature editorial that highlighted the significance of the work. The report "reminds us of the 'potency' of very intense exercise," wrote Edward F. Coyle, a University of Texas physiologist and an expert on human performance. "It appears that this is the first scientific documentation that very intense sprint training in untrained people can markedly increase aerobic endurance. . . . In other words, we are reminded that intense sprint interval training is very time efficient with much 'bang for the buck.'"

That week, someone from McMaster's public relations staff called me up and said, "We have ten newspaper and television reporters who want to interview you." Nothing I'd done to date had ever attracted that sort of attention before. I appeared live on a national morning-news television show. And then the attention snowballed. Ten interview requests multiplied to twenty, and then a hundred. I wasn't able to respond to all the media calls. My in-box overflowed with email messages. It was just surreal. I remember being overwhelmed by the amount of media attention. That single study ended up setting me on the research path that I have followed ever since.

The Second Experiment

After that first experiment, I started incorporating intervals into my own workouts. I was curious to try things out for myself, of course. But it also made sense for me in practical terms. Time was the biggest factor. By harnessing the power of intervals, my workouts required a third of the time they previously had. (Of course, further research would show that intervals could make exercise even more efficient than that.) Although I was only in my thirties at the time, my left knee was showing the degenerative effects of arthritis—the result of a running injury and subsequent arthroscopic surgery when I was twenty-one. So for my own workouts, I favored low-impact activities. My go-to regimen involved cycling on a stationary cycle, an activity that was very conducive to high-intensity interval training.

The sprints turned out to provide me with benefits besides just the time gains. At the end of our first study our subjects reported that they'd never felt better. I experienced a similar effect. After a few interval workouts I felt supercharged. In fact, the effects were so remarkable that I started thinking about how to quantify them. Just how powerful were they?

My team and I designed an experiment to find out. Once

again, we gathered in a small group for a brainstorming session. Our idea was to directly compare high-intensity interval training with more traditional endurance training. We were moving into uncharted territory here. As far as I knew, no one had ever compared a workout involving a few short, hard intervals with a large amount of steady-state continuous exercise.

We decided to compare our sprint-training program with a strenuous regimen of moderate-intensity endurance training based on the typical physical-activity guidelines. We recruited twenty people and divided them into two groups, with five men and five women in each group. They were similar to the subjects in our first study: mainly university students who took part in intramural sports and some habitual exercise but who weren't pursuing any sort of structured training regimen.

One group was put on a quite rigorous endurance training regimen for six weeks. These subjects rode stationary bicycles five days a week for forty to sixty minutes per day. They cycled at an intensity of 65 percent of their maximal aerobic capacity, which is within the moderate range as recommended in the public health guidelines. The pace was enough to get their heart rate elevated and get them sweating.

Now for the interval-training group. They also went on a six-week-long training regimen, but one that required much less work and time. It was modeled after the protocol used in our first study. The subjects began by spending a couple of minutes warming up on the exercise bicycle. Then they performed a thirty-second-long sprint. They rested for four and a half minutes, and then they did another sprint, repeating this four to six times. Instead of training five days per week, they trained three days.

At this point, it's important to grasp just how little exercise the interval-training group did. Their total weekly time spent working out was only one-third that of the endurance training group. And that's counting their rest periods, which required our test subjects to cycle at a lazy pace. Really, they just slowly turned the pedals while recovering from the previous interval.

If you count only the intervals—that is, if you total the time our test subjects were required to perform hard exercise, then our sprinters worked out for just under ten minutes a week. Compare that with the other group's four and a half hours of continuous moderate-intensity exercise per week. The time the sprinters spent exercising amounts to less than 5 percent compared with the endurance group.

Another way to compare the two groups is to consider the amount of energy they expended as they pedaled their bikes. We measured the work in kilojoules (kJ), a unit of energy. Our endurance trainers completed 2,250 kJ in a week of training. That is enough energy to keep a 75-watt lightbulb burning for more than eight hours. Meanwhile, our sprinters completed just a tenth of that. They only did about 225 kJ of work per week—enough energy to keep the bulb burning for about fifty minutes.

How did the results for the two groups compare? Basically,

the improvements were the same for every fitness parameter that we measured. That means both groups improved following the training, but we could not detect any significant differences in the extent of the change between the two groups. The increase in aerobic fitness? The same. The increase in mitochondria in the subjects' muscles? The same. The change in fuel use and, particularly, the subjects' ability to burn fat during exercise? The same.

In short, the experiment showed that approximately ten minutes of hard exercise a week boosted overall fitness to the same extent as four and a half hours per week of traditional endurance training. It's mind-blowing. A tiny bit of sprint training has the same effect on the human body as a whole lot of endurance training—*despite a much lower training volume and time commitment.*

So Is It Possible to Get Fit in Just Minutes a Week?

The answer is an unequivocal yes. Remember those coaches and trainers who thought intervals were only for sprinters? That intervals didn't do much for aerobic conditioning? They were wrong. High-intensity intervals may be the most potent method we've yet discovered to boost cardiorespiratory fitness.

Under strictly controlled conditions, in experiments that have been published in the most reputable peer-reviewed physiology journals, my lab and others around the world have shown that small amounts of interval training can produce benefits we usually associate with large amounts of endurance training.

The harder and faster you go, the less time your exercise requires. Go as hard as you can in short bursts, and you can get the benefits of an endurance exercise regimen with less than 5 percent of the time spent in hard exercise, 10 percent of the work expended, and only one-third of the total workout time commitment. To this day, more than a decade after our first experiment, that fact still boggles my mind.

But if teeth-grinding, face-reddening effort isn't really your thing, you can still get the benefits of interval training. The harder you go, the more efficient your workout will be in terms of time commitment. If sprinting *as fast as you possibly can* is not appealing, or not something you are capable of doing, then you can give 90 percent. Or 80 percent. Even interval *walking* seems to be better than doing the same exercise at a steady-state pace. So long as you vary the intensity, you'll still derive proportionally more benefit from your workout than the traditional moderate approach.

In the next chapter, I'll describe the exciting, exploding *science* of interval training—as well as the current scientific thinking on how and why a little bit of interval training is able to trigger its remarkably beneficial effects.

Then we'll consider the origins of interval training, including a discussion of how some of the greatest athletes in history have employed interval training to conduct world-record-breaking feats.

Then the real fun begins. We'll provide samples of interval

workouts. I'll also discuss the *psychology* of interval training: how to trick your mind and body to blast through the sprints as effectively as possible. And finally, we'll consider the future of exercise and how we might work out in the decades to come. Let's get to it!

CHAPTER TWO

How Intensity Works

NOT EVERYONE WAS CONVINCED BY THE FIRST FEW interval-training studies to come out of our lab. After the first study was published in 2005, we received a lot of incredulous responses from other physiologists. My laboratory group participates in an annual conference in which other scientists in the region get together to discuss their research. I remember that after one of my students had presented our results, another scientist stood up and essentially called our research BS. To paraphrase: "Come on—you're going to sprint for a few minutes and you get these incredible performance benefits? In only two weeks? That's hard to believe."

On one level I could understand the skepticism. It seemed incredible to me, too. But soon after our study came out, we conducted additional experiments that verified our initial results. More editorials in prestigious journals followed. "For those looking to do the least possible work to be fit, the work of Gibala et al. presents hope," Keith Baar wrote in the *Journal of Physiology*. "In as little as 3 min . . . we can improve our VO_{2max} and performance."

Other scientists around the world replicated our results. Interval training was for real. But *why* did it work? How was it possible that such small amounts of exercise could trigger such large benefits? That's the question some of the best exercise scientists in the world were asking one another back in the mid-2000s, as people marveled at the results of our interval training studies. And I was no different. When I'd appear on TV shows or be interviewed in newspapers, the television host or reporter would ask me why intervals had the potency that they did. And I didn't really know. But I knew one thing: I wanted to find out.

REMEMBER HOW I'd just begun my second three-year contract as an assistant professor when my lab started doing the HIIT research? Thanks to that *Journal of Applied Physiology* paper and other studies like it, I was promoted from assistant professor to associate professor with tenure. The 2005–2006 academic year provided the opportunity for my first-ever sabbatical. That's a phenomenon common in academia in which a professor is freed from teaching and administrative commitments to devote a year to research. I could dig into whatever the heck I wanted. To some people that sounds wonderful, but it's a lot of pressure, especially at the beginning of your career. I needed something high-impact to maintain my trajectory.

So I was looking for ways to push my research forward—and to do that, I just followed my intellectual curiosity. Here was my chance to answer the question that kept me up at night: *Why* did interval training work? Why was it so potent?

It was a tricky question to answer because the discipline of exercise physiology still hadn't completely figured out how *traditional* endurance training triggered performance benefits. We knew that training—whether that entailed laps around a track or lots of cycling on the road—triggered adaptations on the entire body that made it possible to exercise faster and longer the next time around.

Some of the most exciting work involved looking at the molecular signals in muscles that were activated by endurance exercise. Scientists were learning that some of those proteins were, in fact, *signals* that prompted the body to change so it could more easily exercise the next time around.

Some of the best research into this process was happening in the laboratory of an Australian colleague named Mark Hargreaves. I wanted to do the same thing Mark was doing at his lab, but rather than studying the effects of endurance training, I wanted to look at how the body responded to *interval* training.

First, I had to pitch Mark on the idea. Conveniently, this was around the time that I started to get invited to conferences held by the Gatorade Sports Science Institute (GSSI). For exercise physiologists in the mid-2000s, the GSSI board was sort of like the top fraternity. Its members were some of the best minds and most respected scientists in the field.

Mark was on the GSSI Science Advisory Board, which would meet every year and invite a few newbies to speak on cutting-edge topics. Getting invited to one of these meetings was a big deal for your career—kind of like getting booked on *The Tonight Show Starring Johnny Carson* if you were a comedian. And then if you got invited to sit on the GSSI Science Advisory Board? That was like Johnny inviting you over after your set to sit in the chair next to him. You'd made it.

One of the first GSSI meetings that I attended took place at the Arizona Biltmore, an historic, beautifully designed hotel in Phoenix. The event went so well that afterward I was invited to join the board. That was the transition. I was in the club.

I believe that also was the meeting where I first asked Mark whether I could work in his Melbourne laboratory. Ever gracious, Mark was receptive to the idea. I soon found myself in Melbourne at the start of my sabbatical. The challenge now was to determine the mechanisms of why interval training was so potent. To stay at the forefront of this research, we had to do it before anyone else did. The race was on.

Physiology 101

To understand what happened next, it's necessary to understand a few things about the physiology of fitness. One important aspect of fitness is the ability of your heart and lungs to pump blood and oxygen throughout your body. This is what most people mean when they talk about "cardio" fitness. Another component of fitness is the ability of your muscles to use the oxygen that gets delivered. Muscles use oxygen to burn fuels, such as sugars and fats, in a complex process that yields the energy-laden molecule known as adenosine triphosphate, or ATP.

Every movement we make requires ATP, and the body has a complex and remarkable system to ensure that we have ATP when we need it. While you're sitting quietly, reading a book about interval training, say, your overall demand for ATP is relatively low. So all the things your body does to get ATP to your muscles tend to be relaxed and slow—including your heart rate, your breathing rate, the amount of oxygen you're distributing to the muscles. But let's say a smoke alarm goes off, there's a fire, and you've got to sprint to safety as fast as you can. Your demand for ATP skyrockets, and so your heart pounds faster, and your breathing gets deeper and harder.

Let's back up a minute and look at what happened to your body the moment you started to sprint from the fire. You heard the smoke alarm, got up off the couch, and started hightailing it out of your house. To allow you to do that, a whole bunch of things happened at once in the body. All of them involved the energy-laden molecule ATP.

Muscles store only a small amount of ATP. It's a relatively heavy molecule and therefore just not efficient to keep on hand in large quantities. Instead of storing large amounts of ATP, your body stores energy in other ways. It's similar to the way you keep your money. You could keep a lot of quarters around to spend, but those tend to be inconveniently heavy; it's more efficient to store the funds in the more compact form of paper currency, in a diverse selection of denominations, such as one-, five-, and twenty-dollar bills. Rather than carrying around huge quantities of coins in your pockets, you keep a wad of bills in various denominations. Then you "convert" the bill when you need to spend the money.

Muscles do something similar with energy. One convenient form of energy is a molecule called phosphocreatine. Think of this molecule as dollar bills in your wallet. Phosphocreatine is easy to convert, but you tend not to keep it in large quantities. It's used to immediately resupply ATP when your muscle starts to contract. Phosphocreatine fuels the lion's share of the energy during your first few seconds of sprinting. This process is extremely fast, but the capacity is very limited.

Another process that can supply ATP relatively quickly is anaerobic glycolysis, which involves the partial breakdown of sugars stored in muscle. You can think of these as the fives in your wallet. The process is quick but rather inefficient, and can get bogged down by the formation of metabolic by-products.

The classic by-product is lactic acid, which accumulates in the muscles of sprinters and is part of the reason why they eventually slow down during a race. The role of lactic acid formation in fatigue is actually quite a controversial topic that could be the focus of a separate chapter or an entire book. For our purposes here, suffice it to say that anaerobic glycolysis is a limited capacity system.

By far, the most efficient way to supply energy is through a process called oxidative metabolism, which involves the use of oxygen to burn fuels such as sugars and fats—the large-denomination bills in your wallet. While slower than the other two processes, oxidative metabolism provides the capacity to utilize many different fuels. The other nice thing about it is that, given adequate fuel availability, the capacity is almost limitless.

Oxidative metabolism really is an ingenious process. It's pretty complex stuff, and one of the scientists who helped figure it out won a Nobel Prize. But what happens in the muscle cell when the oxygen gets in there? The lion's share of this ATP-production process happens in specialized structures in the cell called mitochondria, which suck in oxygen and fuels to generate high-energy ATP molecules. The more mitochondria a cell has, the greater its capacity to produce ATP for energy. For example, spermatozoa are comparatively small cells, but their midpoint is packed with mitochondria, the better to power the tail thrashing that propels the sperm toward the egg.

To summarize, the key to aerobic metabolism is all about getting oxygen and fuel to the mitochondria in the muscles.

This ability largely determines our capacity to perform exercise and, in turn, our overall fitness level. When we exercise regularly, the body gets better at each step in the process. This is the process of training. The challenge for me on my sabbatical was to determine why intervals were so effective at boosting aerobic metabolism. Quite literally, how did the intervals "signal" to the body that it needed to make changes?

Our Response to Exercise

Exercise is traditionally grouped into two broad categories. Endurance exercise typically refers to long-duration, low- to moderate-intensity activity that increases the body's ability to use oxygen to produce energy for sustained movement. Many people think of jogging when they think of this type of exercise, although it also encompasses everything from long-distance swimming to cycling.

The other category features short-duration intense exercise that is usually associated with building muscle strength and size. Many people refer to this type as resistance training, and it encompasses everything from bodyweight push-ups to heavy barbell squats as well as the many exercises that can be done on the pulley-equipped machines found in fitness centers.

What is so fascinating about interval training is that it seems to occupy a middle ground between the two broad categories. It's short duration and high intensity, like strength training, but triggers effects on the body we'd previously associated with endurance training—in a lot less time.

Before we get to why interval training is so effective, let's first consider how the body responds to any type of exercise, which scientists like me consider a *stress* on the body. A physician named Hans Selye in the 1930s developed a theory about the way the body responds when stressed. Selye's general adaptation syndrome says that the body responds in a manner intended to *reduce* the stress the next time we experience it.

When you're at rest, you're in a state called homeostasis. Your heart rate and breathing rate are relatively low and constant, and there is a good match between the body's demand for energy and its capacity to supply it. But once you start exercising, the disturbance to homeostasis throws the body out of whack. The body needs more oxygen than you're giving it. The body responds by increasing heart rate and breathing faster to get more oxygen to the muscles. Then, once the stressor is gone and the body is recovering, the body *adapts* so that the next time the same stressor presents itself, it disturbs the body to a lesser degree.

Take cycling. If you mash your pedals so hard that you struggle for breath, and you do it long enough and frequently enough, your body changes so that the next time you hop on a bike, you don't struggle for breath so much. "Hey!" the body says to itself when it experiences a stressful bout of exercise. "This stuff *hurts*! We need to make some changes to ensure it doesn't hurt so much next time!"

The concept is pretty simple, but the remodeling that takes place is incredibly complex. Repeated over time, exercise provokes a response in the body similar to a major building renovation—except that the normal work going on inside needs to continue during construction. The body's like an airport terminal that is completely modernized while it remains operational. The response to endurance exercise involves changes in every element of the pathway that controls the supply and utilization of oxygen to produce energy.

Over time your heart becomes a better and stronger pump, so that it ejects more blood with each beat. The arteries become more flexible—the better to propel blood through the system. Tiny blood vessels called capillaries grow through muscle tissue to more efficiently deliver the blood-borne oxygen to the muscle fibers. And the muscles grow more mitochondria—the powerhouses that actually use the oxygen to burn the fuel to produce ATP. Highly trained endurance athletes have about twice the mitochondria in their muscles as your average couch potato does.

It's really an amazing process. Think of a garden hose. If your garden hose had the same response to stress as your blood vessels, then it would grow wider and more flexible each time you watered your lawn. Not only that—it might even sprout smaller hoses snaking out from the main conduit, the better to distribute moisture to each individual blade of grass.

So how does the body *know* to trigger all these physiological adaptations? Discovering the precise mechanisms involved has for decades been one of the holy grails of our field. When it comes to muscle cells, scientists believe that certain proteins serve as molecular fuel gauges. Just as a warning light on your car dashboard turns on when the gas in your tank gets low, proteins in your muscles are activated when fuel levels drop. Remember that our most important fuel is ATP. One of the most important fuel-sensing molecules is activated when ATP levels drop.

The energy stored in ATP is a bit like a rechargeable battery. As you exercise, your muscles use up the available energy—the ATP. That ATP gets converted into two metabolic by-products called adenosine diphosphate (ADP) and adenosine monophosphate (AMP). These molecules are like spent batteries that can be recharged. When the body senses a lot of spent batteries lying around, other protein signals get activated.

AMP in particular triggers the activation of a protein with the unwieldy name of 5′-adenosine monophosphate-activated protein kinase, or AMPK. In turn, AMPK activates a protein called—bear with me here, because it, too, is a mouthful—peroxisome proliferator-activated receptor gamma coactivator-1alpha, which most physiologists refer to by its short form, PGC-1α—that last symbol is pronounced "alpha."

PGC-1α turns out to be a pretty special molecule. Some people refer to it as the "master regulator" because of its crucial role in building more mitochondria—important because those additional mitochondria increase the capacity to build more ATP molecules by using oxygen to burn sugars and fats. Some scientists also believe that PGC-1α helps stave off age-related muscle decay. The bottom line for *our* purposes is that PGC-1α is a key signal that triggers skeletal muscle remodeling, allowing the body to perform exercise for longer than it did before.

What's Different About Interval Training

So here's where we are so far: Exercise, the theory goes, uses up ATP, creating lots of AMP, which then turns on a series of signals, including the PGC-1 or master switch. But how much exercise does it take to trigger this process? A 2005 study out of Switzerland suggested that the PGC-1α master switch could be turned on only with contractions that were repeated and sustained *for over an hour at a time.*

That's a lot of roadwork.

And yet we were seeing similar muscle remodeling triggered by just a few minutes of exercise a week. Really hard exercise, admittedly. An effort that required you to cycle as hard as you could. But still!

That year, we submitted a study to the *Journal of Physiology* in which sixteen college-age students performed six training sessions over two weeks. In each training session, half the subjects cycled continuously for 90 to 120 minutes at a moderate-intensity pace. The other eight subjects performed four to six 30-second sprints at an all-out pace, separated by 4 minutes of recovery. The endurance training group's total time commitment was 10.5 hours over the two-week period. In contrast, the total training commitment for the sprint group was about 2.5 hours, including the recovery periods—although the total amount of hard exercise for the sprint group was just 18 minutes.

Once the two weeks were up, testing revealed that the two groups had improved to a virtually identical extent in every measure we tested. Both groups improved their cycling time-trial performance by the same amount and also demonstrated remarkably similar changes in the molecular makeup of their muscles. Considering the difference in the amount of time the two groups trained, it was an incredible result. A total of just 18 minutes of very intense exercise produced the same benefits as 10.5 hours of traditional endurance training.

To physiologists and just about anyone else who followed exercise science, interval training seemed like a miraculous shortcut. But how did it work?

That's what I was trying to figure out in Mark Hargreaves's Melbourne laboratory, where I was working alongside a postdoctoral researcher named Sean McGee, who was an expert at analyzing the molecular changes that exercise prompted.

We took muscle biopsies from subjects who performed a single session of interval training, involving four 30-second all-out cycling efforts, with each burst separated by several minutes of rest. Our analyses revealed two remarkable things.

First, we discovered that a series of short, hard intervals could really ramp up the production of PGC-1α. That is, just a few minutes of sprints had activated the PGC-1α master switch. People were absolutely blown away by this. Remember, some scientists thought that the PGC-1α switch could be turned on only by more than an hour at a time of endurance training. In fact the total amount of exercise done by our subjects was just one-twentieth of that performed in some previous endurance studies—and less than one-third the total training time. We'd shown that this master switch, PGC-1α, could be activated by a lot less total exercise than anyone ever thought possible.

The second important thing that we discovered in the Melbourne lab involved *how* the PGC-1α switch was activated. It turned out that several of the signals believed to crank up PGC-1α after endurance exercise could also be turned on by a few short, hard intervals. For example, one of the proteins activated was our old friend AMPK—the same one that responds to long bouts of endurance-type exercise.

In Melbourne, we established that interval training could activate the same pathways that triggered adaptations to endurance training. Remember the doubters who previously called the results of our initial research into question? The fact that we now had molecular evidence of the potency of interval training silenced them. It legitimized our research. That felt pretty good. It also went a long way toward selling the rest of the field on the power of interval training. We'd not only shown this shortcut to fitness benefits existed; now we'd also taken an important step toward showing how it worked.

Why All This Matters to You

Mysteries persist regarding the physiology of interval training. What regulates the cardiovascular side of things? For example, why does the heart become a better pump, and how do we grow more blood vessels after a few short, hard intervals? My McMaster colleague Maureen MacDonald, a cardiovascular physiologist, and other bright minds across the world are working on these important questions.

I'll tell you what I suspect is going on in the muscle, though. And if what I suspect is true, then it'll go a long way toward helping you choose the sort of workouts that you do.

It all comes back to the concept of fuel gauges. The traditional thinking is that endurance exercise leads to a progressive decrease in ATP and a gradual emptying of the fuel reservoirs needed to replenish ATP. The gradual reduction in fuel stores that happens during prolonged exercise activates the molecular signals that regulate adaptation and muscle remodeling. So for endurance exercise, the longer the exercise bout and the greater the fuel depletion, the larger the adaptive response. Exercise for longer, get more fit.

The situation is different with intervals, however. With a few short sprints of twenty or thirty seconds, the *total* amount of fuel depletion is modest, especially when compared with what can happen over a prolonged period of moderate-intensity continuous exercise. And yet, we have shown that a series of short, hard intervals can activate molecular signaling pathways to the same extent as traditional endurance training does. How can this be?

Here's what I think: With interval exercise, it's the dramatic *rate* at which fuel stores in the muscle are changing. Not the absolute level. It's the *rate* of fuel depletion that's key. Interval exercise is also different from endurance training because more of the muscle is involved in the exercise.

Muscle fibers are generally grouped into two broad categories. Smaller type I fibers, also called slow-twitch fibers, make up about half the overall muscle tissue. These tend to be "recruited" for relatively easy movements that don't require a lot of force. These fibers also are the ones used mainly during moderate-intensity endurance exercise.

Type II muscle fibers, also known as fast-twitch fibers, tend to be recruited for fast, powerful movements that require a lot of force. The effort demanded during high-intensity interval exercise recruits type I muscle fibers *and* the larger type II fibers. Sprinting is hard work, and it takes all the muscle fibers to do it. Because interval training recruits the entirety of the muscle, the muscle uses up available fuel at a much faster rate—so fast, in fact, that even short durations of the most intense flavors of interval training trigger training adaptations. Remember the theory based on Hans Selye's stress-adaptation theory, that the effects of exercise stemmed from the body's disturbance from homeostasis? Interval training may be so effective because it creates big changes from homeostasis—a.k.a. a high degree of stress—in a small amount of time.

Other studies have illustrated the benefits of a technique called exercise snacking—breaking a workout into multiple chunks spread throughout a day. In a 2014 study conducted out of New Zealand's University of Otago, researchers tracked blood sugar in subjects who performed two different exercise interventions. The traditional group performed 30 minutes of continuous exercise once a day. The snacking group performed a quick interval workout before each meal—specifically, six hard reps of minute-long incline walking. The interval workout proved much more effective at reducing subjects' blood sugar, even though the total time spent exercising was the same. Perhaps breaking up exercises into these snacks creates more disturbances in the body's equilibrium. And that's why it's so effective.

The other thing we're learning from interval training studies is that the *size* of the disturbance matters, too. Generally the lesson is, the bigger the disturbance, the better; the bigger the disturbance from homeostasis, the greater the adaptation. So going from homeostasis at rest to a light jog is good. But going from rest to a full run is better. And best of all is going from rest to an all-out sprint. And even better than all that, if you're really looking to cram the biggest performance benefits into the smallest amount of time? Repeat the number of disturbances in a single workout by doing intervals—whether they're light-jogging intervals or as-hard-as-you-can Wingate tests.

In the old endurance-training way of thinking, what was most important was exhausting ourselves through long, slow exercise, which in turn caused a slow and steady absolute decrease in fuel reserves. That suggested it was less about intensity and more about duration—not so much how *hard* we exercised but that we exercised long enough to drain the batteries. But who has infinite free time to exercise?

Now, the new interval-way of thinking suggests that exercise *duration* is a lot less important than exercise *intensity*. The trick is to drop the fuel levels as quickly as possible. Doing it once is great. Doing it more frequently is better. The volatile nature of the stimulus is what's key. Mix it up! Disturb homeostasis! Traditional endurance training sees one main disturbance from homeostasis, right at the beginning. In contrast, interval training gives you as many disturbances as you have repetitions. You harness the power of the disturbance that happens at the start of aerobic exercise, only it's more pronounced. And then you do it again and again and again.

So it's not just the absolute amount of fuel in the cell; it's also the rate at which the stuff is falling. That's important to people who don't have much time. Intervals provide a shortcut. You can drop the fuel gauges really fast by going as hard as possible, particularly if you repeat that a few times. And you'll gain in *minutes* the benefits that once were thought possible only with *hours* of exercise.

CHAPTER THREE

How It All Got Started

THE REALLY REVOLUTIONARY THING ABOUT INTERVAL training is this: It's based on the precept that *intensity* is more important than duration. I'll put that another way. How *long* you work out matters less than how *hard* you work.

Bear in mind, we're talking about developing and maintaining aerobic fitness, which is the ability to exert the body over time. This happens to be the form of fitness that is most important to maintaining a long, healthy, and active life, as well as fighting aging and avoiding many chronic diseases. Things are a little different if your only concern is developing strength, although it is pretty easy to design workouts that simultaneously develop strength and aerobic fitness—as you'll see in later chapters.

The idea that exercise *intensity* is more important than *duration* runs counter to a lot of people's ideas about fitness. It does sound strange—that the most time-efficient way to prepare your body to work at a steady pace for long distances is to go really fast for short spurts of time. But this approach to fitness has been around for nearly a century. It's led to dozens of world records and Olympic gold medals; it's also created lots of scientific debate that's really only been settled in the past ten years or so.

We know for certain that rudimentary interval training—the concept of pushing hard for short distances to develop the capacity to go at a steady pace for long distances—was being used as far back as a century ago, in Finland by the Olympic champion runner Hannes Kolehmainen. He won three gold medals at the 1912 Summer Games in Stockholm, in the 5,000-meter, the 10,000-meter, and the cross-country events. Another Finnish runner, Paavo Nurmi, used interval-training techniques to win nine Olympic gold medals, including five at the 1924 Paris Games, where he won gold in the 1,500-meter and 5,000-meter events despite their being held just fifty-five minutes apart. And the greatest innovator of the next generation of legendary runners, Emil Zátopek, based his training on Nurmi's techniques, to the extent that intervals were virtually all that Zátopek did—and in numbers that seem incredible today.

While training for the 1952 Olympics, Zátopek conducted a

daily workout that involved twenty repeats of a 200-meter sprint, forty repeats of a 400-meter sprint, and then a final twenty repeats of a 200-meter sprint, according to the 1973 edition of Fred Wilt's book *How They Train*. That's a total of almost 15 miles per day in intervals—and Zátopek jogged slowly between the intervals for 200 meters at a time.

So how did Zátopek's training affect his Olympic performance? The Czech runner decided to compete in the marathon event at the 1952 Olympic Games, in addition to the 5,000- and 10,000-meter races—despite never before having run a marathon. He won both middle-distance events, and then came the long-distance event. Here was a man who had trained purely through the use of short sprints entering an event that required running at a fast pace over a distance of 26.2 miles. According to a feature in the *Guardian* newspaper about stunning Olympic moments, which recounted the 1952 marathon, Zátopek felt he lacked the required knowledge to pace himself in such a long event. So he opted to stay near the world record holder, Jim Peters of Britain, who ran the first 10 miles at a record pace. Zátopek could be talkative during his runs. Near the 11-mile mark, he came up behind Peters. "Jim, is this pace too fast?" Zátopek asked. As the more experienced runner, Peters felt the pace was already blindingly fast. But to kid Zátopek, Peters replied that the pace wasn't fast enough. Then Peters watched, astonished, as Zátopek sped up, easily passing the Brit. Soon after, a dispirited Peters was out of gas. He dropped out of the marathon, and Zátopek won the event.

His feat of winning golds in the 5,000-meter, 10,000-meter, and marathon events has never been matched.

Two years after Zátopek's remarkable 1952 Olympics success, another historic performance helped show coaches and athletes the benefits of interval training. At the time, the world record in the mile was held by Sweden's Gunder Hägg, who'd run the event in 1945 in 4:01.4. As recounted in Neal Bascomb's excellent book on the events, *The Perfect Mile*, Roger Bannister was an English student pursuing his medical degree at Oxford University in 1954. Bannister wanted to break not only the record but also the four-minute mile. Since the event tended to be held on a quarter-mile track, such a record would require Bannister to run faster than a minute per lap for the four-lap race. So he trained using intervals—approximately ten quarter-mile sprints—which he started with an average time of 66 seconds a lap and eventually managed to winnow down to a 63-second average.

Bannister set the date of May 6, 1954, to attempt to break the record, according to Bascomb. The venue would be the same cinder track where Bannister had run as an Oxford undergraduate. The day turned out to be windy, and wind resistance can affect the setting of any record. But just minutes before the race was to start, the air stilled, and Bannister set off. He ran the first lap in 58 seconds, the second in 60 seconds, and the third in 63 seconds, meaning he would have to blaze through the final lap in less than 59 seconds. Which he subsequently did, making the mile in 3:59.4, a full 2 seconds faster than the previous world record, thanks in part to his desire and his willpower, and thanks also to the power of intervals.

Bannister, Zátopek, Nurmi—each of them harnessed the power of intervals, and the principal of intensity over duration, to conduct remarkable and record-breaking athletic feats, as did many other coaches and athletes of the era, including such Americans as the University of Kansas track coach Homer Woodson "Bill" Hargiss and the University of Oregon's Bill Bowerman, the cofounder of Nike.

Which brings up a puzzle. If these runners and coaches were savvy enough about the benefits of intervals in the 1920s, '30s, '40s, and '50s, then why'd it take so long for the rest of us to catch up? Mainstream culture has seen plenty of fitness fads come and go—jogging in the '70s, Jane Fonda's aerobics in the '80s, fitness boot camps in the '90s. So why did we have to wait until the past decade or so for the time-saving benefits of high-intensity interval training to cross over from athletes to the mainstream? Why'd it take so long to go from the track to our living rooms and fitness studios?

The short answer is that interval training *did* cross over—just not in the way we expected. It started in the late '50s and early '60s in one of the least likely places imaginable: the frozen tundra of the Canadian Arctic. This was the height of the cold war, when democracy's champion, the United States of America, faced off against Communist Russia in a tactical game of global chess played with nuclear warheads. If Russia were ever to stage a surprise attack against the United States, the shortest route went over the top of the world—northern Canada. So America's ally, Canada, stationed pilots and planes in remote outposts in the Great White North. The Canucks were supposed to be ready at a moment's notice to scramble into the air, to combat the Russian bombers many people expected to fly in from Siberian air force bases.

The problem was, many of these pilots were stuck in places where in wintertime the sun barely rose, and the average temperature barely ever climbed past zero Fahrenheit. For months at a time. These guys were supposed to be protecting the fate of the free world. Instead, many lost their fitness because they weren't getting enough daily activity. By the mid-1950s, fully one-third of Canadian pilots were so out of shape they were considered unfit to fly.

So the Royal Canadian Air Force decided to do something about it. Remember, this was starting in the 1950s, when the academic discipline of exercise science barely existed. But luckily Canada's Department of National Defence had one of the first exercise scientists working for it. His name was Bill Orban, and he was a kid from the Canadian prairies who'd attended the University of California, Berkeley, on a hockey scholarship and eventually got his PhD from the University of Illinois.

Orban was the perfect person for the job. As a hockey player from Saskatchewan, he was accustomed to the challenge of staying fit in areas with long, dark winters. Another thing: he'd noticed that hockey players' short bursts of on-ice activity were amazing at providing aerobic fitness. Finally, while at the University of Illinois, he observed that one could exercise for long periods at low intensity without necessarily getting any fitter. Putting all this together, he concluded that the speed and effort expended during a workout were more important than the overall length of time one exercised.

Orban's task was tough. He had to keep these Canadian airmen in fighting shape despite the fact that they lived in little warm boxes in remote outposts in the Great White North. The workout would have to be conducted under strict limitations. It had to require nothing in the way of specialized fitness equipment. What's more, it had to keep its practitioners in fit, fighting shape—and it had to keep its practitioners strong.

The protocol that Orban designed was ingenious. He called it 5BX, for "five basic exercises," and he set it into a series of progressively more difficult workouts designed to be repeated three times a week. The Royal Canadian Air Force tested and refined the 5BX protocol for three years. Orban designed the entry level for people who had never before worked out. The hardest level was only for "champion athletes"—sports pros and Olympians. The second-tier level begins with standing toe-touch stretches; progresses to sit-ups, back extensions, and push-ups; and ends with the longest bout of exercise: running in place for a beginning duration of 335 paces.

The best part? Because it relied on all-out intensity, the workout lasted a total of just eleven minutes. Critics in the Department of National Defence wanted the program to include longer bouts of exercise, but Orban stood his ground. You didn't need long bouts of continuous exercise to improve and maintain fitness, he said. It was the *intensity* that was most important.

That's right—the Royal Canadian Air Force prescribed high-intensity interval training to its airmen back in the late 1950s. Soon, the workout spread outside the military. The RCAF published a pamphlet on 5BX that was available to the general public in 1961. The first printing of sixteen thousand copies sold out quickly, so the RCAF printed it again, in a higher quantity, and that one sold out, too. By 1963, after many other printings, some of which included a similar fitness regimen for women, known as the XBX, the number of pamphlets sold totaled 5.8 million copies, according to an excellent account of the 5BX phenomenon by journalist Alex Hutchinson, published in the newspaper *Globe and Mail*. (Hutchinson's article is the source of many of the 5BX facts I mention.) Today, more than twenty-three million copies of the 5BX workout have been published. It is available for free on the Internet—just search for "5BX" to check it out. The comedian George Burns, who lived to be one hundred, credited 5BX with his longevity and health. Britain's Prince William has been an enthusiastic practitioner. And in 2014, according to the *Telegraph*, Dame Helen Mirren cited it as the "elixir of her youth and the reason why she can still rock a coral bikini in her sixties."

Today, certain elements of the 5BX aren't encouraged. The

sit-ups suggested in some versions of the protocol are said to be bad for the back, for example. But the Royal Canadian Air Force's idea of a short, do-anywhere workout to complement both aerobic fitness and physical strength persists in a wide range of regimens today. "P90X is just 5BX without the marketing," observed Carl Foster, the American College of Sports Medicine's former president. Similarly, exercise physiologist Chris Jordan, of the Johnson & Johnson Human Performance Institute in Orlando, Florida, published in ACSM's *Health and Fitness Journal* a high-intensity circuit-training program designed to achieve similar aims as 5BX's. "The combination of aerobic and resistance training in a high-intensity, limited-rest design can deliver numerous health benefits in much less time than traditional programs," Jordan and his coauthor, Brett Klika, wrote.

According to Hutchinson's *Globe and Mail* article, Jordan knew about 5BX from his work as a physiologist for the British Army in the 1990s; later, while working for the US Air Force, Jordan designed another version for American pilots. The *New York Times'* Well Blog subsequently published, in May 2013, the Jordan program under the heading "The Scientific Seven-Minute Workout," helping the program achieve 5BX-levels of popularity, and a software program based on Jordan's program became one of the most downloaded fitness apps.

AND YET, aside from the strange anomaly of a fitness fad that came from the Great White North, the question remains: Why, if athletes knew how potent intervals were in the 1920s, '30s,

'40s, and '50s, didn't more of us also grasp how efficient the stuff was as a form of health-promoting exercise?

It comes down to this: The science had to play catch-up. The discipline of exercise physiology was still relatively young in the mid-twentieth century. We hadn't quite grasped that the same things that boosted athletic performance also improved human health. In those early years, the field was mainly concerned with describing basic physiological responses to exercise: things like the circulatory response to prolonged activity and what effect dehydration had on work capacity. It took some time for enterprising early physiologists to study topics like the most time-efficient training mechanisms. Besides, competitive track athletes aren't the most transparent when it comes to sharing their training secrets. After the records are set, there's typically a lag of years or even decades before the winners disclose the training regimens that allowed them to establish global bests. It's also important to realize that training practices happen in waves. Intervals were the order of the day for runners in the '50s; later, long, slow running and the practice of tapering—decreasing training volume in the weeks leading up to a competition—generated more interest.

Interval training did have its champions and early adopters, who argued that intense bursts of exercise were an effective way to boost athletic performance. Woldemar Gerschler, a professor of physical education and a track coach at Freiburg University in southwest Germany, is generally regarded as the pioneer who first examined the physiological effects of interval training in a disciplined and scientific manner, beginning in the late 1930s. According to a column by track coach Peter Thompson in *Athletics Weekly*, Gerschler's interval training helped the German athlete Rudolf Harbig run 800 meters in 1:46.6—beating the previous world record by 1.6 seconds. The record would stand for a remarkable sixteen years, according to Thompson, and the next athlete to break it, Roger Moens, was also coached by Gerschler. Even more astonishing, a month after he set the 800-meter record, Harbig ran the 400-meter dash in another record-setting time of 46 seconds flat.

Thanks to such results, Gerschler became a much-quoted presence in English-language track magazines over the subsequent decades and coauthored a German-language book on the topic, *Das Intervall-training*. In 1959 Gerschler's collaborators, cardiologists Hans Reindell and Helmut Roskamm, published the first description of interval training in a scientific journal.

Gerschler regarded interval training as superior to standard, steady-state endurance training because it improved performance more quickly. "The long distance run, slow and uninterrupted, requires training over exceptionally long distances to obtain the necessary powerful stimulus," he wrote in a 1963 article for the newsletter *Track Technique*. "This would require running day after day, for hours, which would tire one more for its monotony than by the required efforts." The following year, in an interview for the same publication, he said, "Our observations taught us that long distance running with very little strain on the individual is close to worthless." Gerschler's writing reflects a rivalry between interval training and traditional endurance-running methods that exists to this day: "Long distance running for the purpose of increasing the ability to run faster over a middle distance race is surely of no importance."

In the United States, it was Edward L. Fox whose evangelism for intervals came closest to pushing the training method into something the mainstream considered as health-promoting fitness. A professor at Ohio State University, Fox became interested in the science of interval training after he'd been struck by its potency while conducting research for the US military. In the early 1960s, folks in Washington worried that Communist China was going to attack India, America's cold war ally, by invading over the Himalayas, the world's highest mountain range. Washington wanted to know how US soldiers would fare if thousands of them were suddenly parachuted into the high-altitude peaks and valleys that dominated the Indian-Chinese border. So they hired Fox and his academic mentor at Ohio State, Donald K. Mathews, to devise an experiment.

Mathews, Fox, and several other Ohio State scientists conducted their research in 1963. As recalled by Dick Bowers of Bowling Green University, who helped conduct the study, the scientists split a number of soldiers into two groups. One group conducted about an hour of interval training per day for eight to ten weeks, while the second conducted two to three hours of conventional Army physical training, such as calisthenics and fast-paced marching. At the end, the scientists transported the troops into California's Sierra Nevada, where they spent three weeks undergoing tests in such things as rifle marksmanship. As sometimes happens in science, the most important thing about this study had little to do with the intended result. It turned out that fitness provided little protection from altitude sickness, which was the topic that really interested the US Army. But the scientists were astonished by something they noticed after the preliminary training, according to Bowers. Despite the interval group's having spent much less time training than the steady-state-exercise group, all soldiers' fitness levels ended up about the same.

The results intrigued the Army enough that they commissioned Mathews and Fox, among others, to conduct further studies on the benefits of interval training. Perhaps, the Army thought, the technique could more quickly whip soft, civilian recruits into fighting shape. Perhaps rather than sending the grunts on those ten-mile runs in their steel-toed boots, the drill sergeants of basic training should have been pushing their charges through series of all-out sprints.

"It appears," Mathews, Fox, and their coauthors concluded after the second study, "that an interval training program which emphasizes short distance running may be expected to produce maximum improvement in cardiovascular endurance as a function of time expended on training." Elsewhere, Mathews and Fox suggested that "the biggest advantage of the interval program over the Army program is the relatively modest amount of time required per session to bring about the desired results."

More than a half-century ago, in other words, the researchers understood that interval training was the fastest and most potent way to boost fitness in a time-efficient manner. Fox and Mathews coauthored the 1974 book *Interval Training*, which provides a conditioning program "for sports and general fitness." The textbook asserts: "Interval training is the supreme way to condition a person."

Mathews and Fox were such ardent believers in the method's power that they credited the technique with spurring a sharp historical decrease in world records for the 100-meter, 400-meter, 1,500-meter, and marathon races since 1900. "As a matter of fact, man should do all physical work in intervals, rather than continuously," they wrote. "For the coach and athlete, interval training programs should form the basis of conditioning for *all sports*," including linebackers, sprinters, shot-putters, and "yes, even the business executive." In fact—perhaps allowing their enthusiasm for the approach to get the better of them—the men advocated an interval approach for everything from shoveling snow to vacuuming the living room to "spading the garden."

"Hopefully, you . . . will consider carefully the [interval training] approach," Fox concludes in the introduction. "Interval training is the most successful and least painful approach to fitness we know."

Fox was right. But few listened. The reason? Scientists of that era disagreed whether interval training was better than endurance training. One of the biggest names to weigh in on the topic was the Scandinavian physiologist, Bengt Saltin, who participated in some of the first academic papers to have been written in English on the unique properties of interval training.

Saltin also happens to have been a mentor of mine. He was the first president of the European College of Sport Science and the author of almost five hundred publications in scientific journals—including five relatively obscure papers that I coauthored with him. I feel privileged to have spent two years working in his laboratory as one of his postdoctoral researchers beginning in 1997. By that point, Saltin was one of the most-respected names in the field of physiology and the director of the Copenhagen Muscle Research Centre at the University of Copenhagen. He struck me as a renaissance man. He was curious about everything.

When I started doing research for this chapter, I had a vague appreciation that Bengt had been interested in interval training early in his career. I certainly did not know that, in fact, he'd acted as a research subject and coauthor on one of the first studies to be published on the subject in a peer-reviewed journal. I was also struck by the way he influenced the debate in the 1970s, when physiologists were divided over whether interval or continuous training was superior. "Several authors are convinced that intermittent training results in the best possible improvement in endurance," Bengt wrote in 1976, using an early synonym for "interval training." Yet he concluded, "interval training does not appear to have an advantage over continuous training in enhancing endurance capacity."

That's right—early on, my mentor concluded that interval training was only about as effective as continuous training. The opinions of renowned experts like Bengt carry a lot of weight and have contributed to a certain viewpoint in the scientific community. People like Edward L. Fox claimed that interval training was a remarkably efficient way to elicit physiological and performance adaptations—a claim that my own studies and those of many others in the physiology community would validate in coming decades. But back in the '70s, the prevailing attitude was that continuous and interval training were about equal. That is, one was no better than the other when matched for the total time people spent exercising. Scientists like Fox, with their suggestions that the Army should be putting its soldiers through sprints in basic training, rather than time-intensive forced marches, were considered outliers.

One influential study, cited by Bengt and others, was led by Duane O. Eddy of Indiana's Ball State University. The researchers took two groups of university-age subjects and put them through two different training programs. The first group was required to pedal continuously at a steady state on a stationary bicycle, at a moderate pace of 70 percent of their VO_{2max}. The other group performed intervals—in a minute-on, minute-off format, alternating sprints at 100 percent of their VO_{2max} with easy cycling. The training was calibrated so that the two groups burned about the same amount of calories during the two types of exercise. At the end of seven weeks of training, the researchers assessed the two groups' fitness. The results were nearly identical. Based on the results of Eddy's study, Bengt concluded in 1976 that, "interval training does not appear to have an advantage over continuous training in enhancing endurance capacity."

How could that be? After all, my own studies, as well as dozens of others conducted by my peers in the past ten years, have concluded that interval training is a far more time-efficient way to boost fitness and health. Some protocols are seemingly more effective by an *order of magnitude*.

So what's up with the study by Eddy and his colleagues? I believe the issue relates to the intensity of the intervals the researchers employed. You think about Paavo Nurmi, Roger Bannister, and Emil Zátopek—many of the intervals they were doing were all-out. The same goes for many of the intervals we conduct in my lab. The sprint pace demanded of our subjects is two or three times higher than the workload they achieve at VO_{2max}.

In contrast, the intervals that Eddy and his fellow scientists asked their subjects to conduct were comparatively easy—at a pace equivalent to 100 percent of their VO_{2max}.

Eddy's intervals weren't hard enough. The influential mid-1970s study that caused Bengt Saltin and, I suspect, many other physiologists to conclude that interval training was only about as powerful as continuous training turns out to have been based on a training program that wasn't intense enough.

That's why interval training didn't cross over into the mainstream in the '70s.

The scientific community's biggest evangelist for interval training, Ohio State University's Ed Fox, died suddenly at the age of forty-four in 1983. What would have happened had Fox lived to conduct further studies into the benefits of interval training? Or if Eddy and his fellow researchers had asked for more from their subjects—for example, for them to perform all-out sprints? Would the course of history have changed?

The public really didn't begin to wake up to the benefits of interval training until 1996, when Izumi Tabata, the Japanese physiologist and an assistant coach of Japan's Olympic speed-skating team, published a study in the journal *Medicine and Science in Sports and Exercise*. The study helped establish just how potent intervals could be in boosting fitness. It was based on a workout devised by the Japanese speed-skating team's coach, Irisawa Koichi, that went like this: After a warm-up of ten minutes, the skaters did eight sets of all-out sprinting for just twenty seconds per interval, separated by rest intervals of ten seconds, for a total of eight sprints—and just two minutes and forty seconds of hard exercise. Notably, Tabata asked his subjects to perform their sprints at an intensity corresponding to 170 percent of VO_{2max}.

The skaters conducted the protocol on a stationary bike four times a week. On a fifth day, the subjects cycled for thirty minutes at a moderate pace and conducted four repeats of the twenty-second sprints. Despite how little hard exercise the subjects performed, they boosted their VO_{2max} by 15 percent. That's quite an improvement considering the relatively small time investment.

Tabata's unconventional protocol percolated through the personal training world. A little under a decade later, I published my own series of studies, most of which required subjects to conduct sprints at an intensity that corresponded to more than 200 percent of their VO_{2max}.

The studies in my laboratory included invasive measurements like muscle biopsies, which shed light on the cellular mechanisms behind the effectiveness of all-out sprints. We also made direct comparisons between the effectiveness of all-out sprinting and traditional endurance training that underscored what we were only beginning to grasp fully: brief intervals, particularly those performed in an all-out manner, were remarkably potent for boosting fitness. The publicity around my studies seemed to help laypeople grasp the concept that intervals could be a way to get fit—fast.

I never had the chance to really question my mentor, Bengt Saltin, about the conclusions he drew in 1976. He died in September 2014 at the age of seventy-nine; I didn't start researching this book until a few months later. What I do know for certain is that he eventually came around to the power of interval training. He followed my career and my publications, as he did those of all his students. The last time I saw him was September 2013. I was in Copenhagen for a scientific meeting, and Bengt and I went out for coffee at a little café just across the street from the University Hospital, where his lab was. The two of us chatted about our usual topics—science, life, our families, and common friends. At one point he mentioned the attention my studies were getting. It was great to see, he said. The work that I was doing was important, he said—and he was proud of me.

And yes, that did make me feel awesome.

CHAPTER FOUR

Beyond Simple Fitness

BY NOW YOU GET THAT INTERVAL TRAINING IS EFFECTIVE. But so far we've mainly focused on training regimens that increase athletic performance. This chapter is about the effect of intervals on *health*.

I think this is the most exciting chapter in the book, and here's why. If you ask most people why they don't exercise, they come back at you with a couple of standard excuses. Absolutely by far the most common excuse is "I don't have time." Getting fit just seems so overwhelming to them. They envision fitness as this thing that exists on the other side of a vast chasm. We're talking the Grand Canyon here. The people are on one side, feeling tired and out of shape. And on the other side is fitness, and all the things that come with it, like more energy, a brighter outlook, a healthier body, and a longer life.

To cross that chasm—to get fit—people have all sorts of perceptions about how much they think they're going to have to work out. They visualize months and months of exercising for hours at a time each week. The thought of even *starting* to get fit is overwhelming.

But they're thinking about the traditional kind of exercise. The slow-and-steady approach. They don't realize that a new and scientifically proven method exists *for them*. Thanks to interval training, we know that you can boost your fitness fast. As few as six sessions of interval training over two weeks triggers physiological adaptations that can *double* your endurance capacity. Six sessions, totaling a single hour's worth of hard exercise, can lower blood sugar in people with diabetes. Six sessions can get you feeling like you're in shape. And if you can tolerate it, a minute's worth of maximal exercise, in the form of three all-out sprints for twenty seconds each can change your physiology as much as fifty minutes of cycling at a moderate pace. Studies from my lab have demonstrated all these things.

The naysayers warn that high-intensity intervals are only for people who are really fit and really motivated. But those naysayers are wrong. Listen: *Some* people shouldn't perform interval training. But it's a rather limited group, and many more—even those with chronic diseases—can benefit from an interval-based approach to fitness.

The Risk-Benefit Analysis

Before we get to the health benefits, I want to discuss something that invariably comes up when I talk about interval training, whether it be at conferences or social gatherings. "Wait a second," people will say to me. "Interval training sounds great. But isn't it risky?"

Here's what I say. First, I point out that not all forms of interval training involve super-intense exercise. For the vast majority of people, any exercise is safer than doing nothing. "The biggest risk is not getting off your couch," says my colleague Maureen MacDonald.

Everything involves some risk. Leaving your house in the morning is incrementally riskier than staying inside. Getting on a plane to head off on a week's vacation is incrementally riskier than staying at home. And heading out on that sailboat is incrementally riskier than sitting inside your room at the resort all day. We do all those things because they have benefits that justify the incrementally increased risk. It's the same for exercise.

According to a 2015 position paper put out by the American College of Sports Medicine, "vigorous-intensity exercise does have a small but measurable acute risk of [cardiovascular disease] complications." That means that for a short period during and soon after the vigorous exercise, there is an elevated risk of a cardiac event, like a heart attack.

In 2016 the American College of Cardiology's Sports and Exercise Leadership Council cited a study that showed the risk for a sudden cardiac death related to vigorous exercise was low, at one per 1.42 million hours of exercise—but nonetheless, that's 17 percent higher than during periods of low or no physical activity. The cardiologists explained that conducting *more* bouts of vigorous exercise decreased the risk of experiencing a heart attack *during* that vigorous exercise. Whether people did interval exercise or steady-state endurance-based training, those who pushed the intensity into vigorous levels five times a week had a "markedly lower" risk of a vigorous-exercise-caused heart attack than those who didn't do any vigorous exercise at all. "Such results," the cardiologists noted, "demonstrate that vigorous physical activity transiently increases the risk for acute cardiac events, *but reduces the overall risk*."

That added emphasis is mine. This is what's so tricky about understanding the risks associated with interval training, and in particular the most intense forms like sprint training. In older people, the risk of a sudden cardiac death during or soon after exercise goes up the more vigorously one exercises. But vigorous exercise makes you more fit, which decreases your overall risk of dying.

To get a better sense of the risk, I talked to one of the authors of that cardiology paper—Paul Thompson, the director of cardiology at Hartford Hospital in Connecticut and a member of the Sports and Exercise Leadership Council. Thompson is a big believer in exercise. He ran to and from work every day for years and has completed twenty-seven Boston Marathons. "If you're totally healthy, there's no increased risk from the high-intensity stuff," he says. "The problem is, if you're in your forties, fifties, and sixties, you may not know whether you have things like atherosclerosis. You don't know if you have cholesterol in your coronary artery. Most sudden heart attacks happen in people with no symptoms."

"If you gave me a thousand people and had one group doing high-intensity exercise and one moderate, the high-intensity group would be more likely to have a sudden heart attack during exercise than the moderate-intensity group," says Thompson.

The key there is *during exercise*. Because things change if you're considering one's risk of heart attack in general. "Both the high intensity and the moderate group are going to live longer and have fewer heart attacks than the sedentary folks," he says.

Do you see the difference? Work out hard and you'll have a slightly higher risk of experiencing a cardiovascular event while you're exercising. But all the rest of the time, that hard exercise is *lowering* your overall risk of developing heart disease. There's a slightly increased short-term risk, as well as a long-term increased benefit. That trade-off is worth it in my eyes, and the time-efficiency of intervals is why I have trained that way for years. However, Thompson points out that the risk of a sudden cardiovascular event increases with age. "If you said to me, 'I'm either going to do nothing or HIIT,'" says Thompson, "I'd say, do HIIT. If you said to me, 'I'm either going to do moderate or HIIT and I'm sixty-five *and* time isn't an issue,' then I'd say, do moderate."

In the decision to perform interval training, as with many

things in life, we're managing risk. The takeaway from all this is that your risk depends on your age. Hard exercise is generally safe for the young and the healthy. But you're going to reach an age at which you'll have to ask yourself whether the benefits of time-effective exercise are worth it. There may be an age when you start to ease off on the really vigorous stuff. And if you've been including interval training in your workout regimen, the age that you have to make that decision will likely be later than it is for those who haven't done any interval training.

Interval Training and Health

Now let's get back to the irrefutable fact that exercise boosts health. First, we're going to discuss in general terms how we know exercise boosts health, and then we'll focus on interval training and the evidence that suggests it has some peculiarly potent health-boosting effects.

The physiological changes that come with exercise and enable us to run faster and harder in the short term, also help us live longer and more active lives—with less chronic disease. This is something that many had suspected: Hippocrates, the ancient Greek physician credited with founding the profession of medicine, for example, said that "eating alone will not keep a man well; he must also take exercise." But it was well into the twentieth century before the link between physical activity and health was scientifically established. In the late 1940s, the British physician Jeremy Morris set out to study the relative health of the approximately thirty-one thousand transit workers who staffed London's bus network. Two broad job classes existed: the *drivers* piloted the crowded double-decker buses around London's congested streets, while the *conductors* moved among the vehicle's passengers, going up and down the stairs to take tickets and maintain order.

The drivers sat for 90 percent of the time they were working. In contrast, the conductors climbed an average of six hundred stairs each shift. In what he called the London Transport Workers Study, Morris examined the employees' medical records and tallied up incidences of heart disease among them. And he discovered a remarkable difference. The more physically active conductors experienced fewer than half the heart attacks than the drivers did and contracted heart disease much less frequently. And when they did experience a cardiac event, the conductors were much more likely to survive it than were the more sedentary drivers.

In 1953 Morris published a classic epidemiological study in the *Lancet* that was one of the first attempts to link activity level to disease risk in a large population. Since then, dozens of epidemiological studies have established that exercise is associated with a reduced risk of developing cardiovascular and many other diseases. It also reduces mortality or the risk of dying from all causes. Today we know that exercising regularly is probably the single most effective thing we can do to prolong life and improve health. Which is why we have guidelines from agencies like the American College of Sports Medicine and the World Health Organization that advocate

150 minutes of weekly moderate-intensity aerobic physical activity to promote health.

Unlike some other public health agencies, the ACSM and the WHO acknowledge a potential role for exercise intensity—their guidelines state that people can opt for 75 minutes of *vigorous*-intensity activity in lieu of 150 minutes of moderate-intensity exercise. Studies out of my lab and others suggest that if you crank up the intensity past vigorous to near maximum or all-out, you can boost health in even less time.

Here, then, are the most important ways that interval training has been shown to increase the quality and quantity of our time on this planet.

Intervals Fight Cardiovascular Disease

Cardiorespiratory fitness is an umbrella term that refers to the health of the heart, lungs, and blood vessels. As I noted earlier, the gold standard measurement of cardiorespiratory fitness is maximal oxygen uptake, or VO_{2max}, which is the maximum rate at which oxygen is used by the body during heavy exercise. (It's sometimes called peak oxygen uptake, or VO_{2peak}. While there are technical differences between the two terms, for most purposes they're synonymous.)

Scientists have realized that VO_{2max} also turns out to be an important indicator of a person's overall health—a kind of "magic number" that goes a long way toward predicting how long you'll live and your risk of developing many chronic diseases. It's probably the most important metric you have.

A lot of factors dictate your VO_{2max}, some of which you can control (like your activity level and whether you smoke), and others that you can't (such as your age and sex). The single most important factor that determines individual differences in VO_{2max} is the amount of blood your heart can pump.

Scientists figured this out in part through comparative physiology—studying physiological differences in other animals besides humans. Especially when comparing animals of similar size but with different aerobic capacities, they found out that the bigger an animal's heart, the higher its VO_{2max}. For example, a dog has more than twice the VO_{2max} of a goat—and a heart that's twice the size. Similarly, a racehorse has more than twice the VO_{2max} of its mammalian cousin, the steer—and again, the racehorse's heart pumps twice as much blood as that of a steer.

The *human* heart is pretty remarkable. In a typical human at rest, the heart can pump about ten pints of blood every minute, regardless of whether the person is fit or a couch potato. Think about that—that's five 32-ounce Big Gulps. Going through your heart each minute.

But it's when a person really gets active that fit hearts distinguish themselves. Thanks to my interval training, I'm in decent shape for a guy my age—hardly an elite athlete, but on the upper end of the spectrum. When I'm really cranking—mashing the pedals of my bike as I climb up the Niagara Escarpment near my house, for example—my heart pumps about five times as much blood as it does when I'm at rest. And for highly trained endurance athletes, like the US Olympic swimming champion Michael Phelps? When *they're* exercising, their hearts can pump up to *eight* times as much blood per minute as they do when at rest—a total of about ten gallons of blood per minute. That's the equivalent of forty Big Gulps. Pumped through the heart. In a single minute.

Here's another way to think about it. Some smaller cars have fuel tanks that hold about ten gallons of gas. At the filling station, it takes more than a minute to fill up these tanks. So the amount of blood that can come out of the human heart during exercise is greater than what comes out of a gas station fuel pump. It's just one more indication of how remarkable the human body is.

So what's the most time-effective way to boost the capacity of the heart to pump blood, which in turn boosts VO_{2max} and reduces the risk of cardiovascular disease and overall mortality? Studies from my lab, and a substantial body of evidence from academics across the world, suggest that it's the more intense forms of interval training, like high-intensity interval training and sprint interval training. "It appears that HIIT improves cardiorespiratory fitness more than [moderate-intensity continuous training does] across a broad range of populations, including healthy sedentary and heart failure patients," states a recent overview in the academic journal *Heart*.

We'll get to those heart-failure patients in a bit, but first let's talk about the way an increase in heart and lung health translates into longer lifespans. The author of the *Heart* article, Matthew Wilson, cited a 2013 study published in the *European Heart Journal* that tracked the health of all French cyclists who had participated in the Tour de France cycling race between 1947 and 2012. The study compared the mortality risk of those 786 cyclists with that of the general population of French adult males. The cyclists had a 41 percent reduction in all-cause mortality risk compared with the community's average risk. Put another way, those French cyclists tended to live 6.3 years longer than the average French male. A similar study comparing runners with an average control group in the United States, published in 2008 in the *Archives of Internal Medicine*, demonstrated even greater effects due to exercise, with 15 percent of the runners having died nineteen years after the study began, compared with 34 percent of the average control group.

Why is that? What is it about exercise that so extended the athletes' lives?

Wilson points out that exercise can decrease the resting heart rate by 20 beats per minute and increase the volume of blood the heart can pump by about 20 percent. That's because, under the influence of physical activity, in normal people the heart's four pumping chambers get bigger. The heart muscle itself gets stronger and better able to pump. The blood vessels also change as a person becomes more fit. There's an improvement in something scientists refer to as "endothelial function." That means the walls of the vessels become more pliable. When we transition from rest to some sort of physical activity, the interior diameter of our blood vessels increases, which permits a greater flow of blood. A 2015 review of scientific studies out of the University of Queensland in Australia revealed that one of the most-used HIIT protocols was able to improve a key measure of endothelial function by twice as much as traditional moderate continuous exercise. In the short term, these sorts of changes mean the heart doesn't have to work as hard to get blood to the parts of the body that require it. Over the long term, it decreases the likelihood of the sort of events that tend to kill us, such as heart attacks and strokes.

About the very long term, though: Getting older tends to change the cardiovascular system in the *opposite* way that becoming fit does. Over time, the heart grows less able to pump, and the vessels grow stiffer, diminishing their capacity to carry blood. The nice thing is that we can substantially slow, and in some cases even reverse, this age-related decline. The way to do this is through exercise. But what sort?

That's the question German cardiovascular physiologist Katharina Meyer asked early in the 1990s. Meyer's academic lineage goes back to the very beginning of research into interval training's effects on the human body. She has been the lead author on studies supervised by Helmut Roskamm, the researcher who coauthored the very first paper on interval training published in an academic journal, back in 1959.

In the late 1990s Meyer conducted a pioneering series of interval-training studies on patients with severe cardiovascular issues—everything from heart failure to those who had undergone bypass surgery. The experiments she conducted were way ahead of their time. "Katharina Meyer was doing things back then that nobody else was doing," recalls Jeff Christle, a Stanford University physiologist who specializes in exercise.

Several of Meyer's most innovative experiments applied interval training to patients who had chronic heart failure, a condition in which the heart does not pump blood as well as it should. The condition can develop after the heart becomes damaged or weakened by other conditions or diseases, including heart attacks, and it can leave patients winded by actions that most people wouldn't consider taxing, such as walking up a slight incline. The poor pumping capacity of the heart often causes fluid to back up in other tissues, including the lungs, leaving people feeling tired and out of breath. In one study, published in 1997, Meyer showed that three weeks of an interval-training protocol improved cardiac function in patients with heart failure, such that cardiorespiratory fitness increased by about 20 percent.

People thought she was crazy. Back then, intense exercise was thought to be dangerous for anyone with any sort of heart issue. Norwegian physiologist Ulrik Wisløff recently led an effort to compile a short history of cardiac rehabilitation. According to Wisløff and his coauthors, well into the middle of the twentieth century, cardiologists prescribed patients who had survived major cardiac events, such as heart attacks, as close to complete bed rest as possible. Essentially, the patients were supposed to lie flat on their backs, on a bed, for thirty days. No activity whatsoever was allowed because the rest was thought to heal the scarring caused by their heart attacks.

Then, in 1952, a pair named Samuel Levine and Bernard Lown suggested that maybe patients would do better if they weren't *completely* immobilized. If, say, the recovering patients sat up in an armchair. Soon, other cardiologists and scientists were suggesting the patients might be able to do even more activity—that actually, up to a point, the *more* active the cardiac patients were, the better.

Remember the early buzz around high-intensity exercise to boost athletic performance propagated by people like Ed Fox in the 1970s? Well, it turns out the world of cardiac rehab experienced a similar phenomenon. Beginning in the '70s, some cardiologists and physiologists felt that the low-intensity exercise they were prescribing their patients wasn't strenuous enough. They felt that cardiac rehab patients could safely engage in activity that was a lot more strenuous. Not only that—the patients *should* work themselves harder. It would be good for them, because the higher activity would assist with the healing of the heart.

Some of this activity included interval training. Scientists who conducted early studies on interval training's effects on cardiac patients included Vojin Smodlaka in New York and Roy Shephard in Toronto. The first of them, Smodlaka, asked some patients to cycle at high intensities for sixty seconds at a time, followed by thirty seconds of rest—and repeat. Afterward, Smodlaka discovered that the cardiac patients who had completed the interval-training program could exercise for twice as long as those who had engaged in continuous training.

A Toronto cardiologist named Terry Kavanagh even convinced eight of his patients, all of whom had experienced heart attacks anywhere from one to four years previously, to train to run the 1973 Boston Marathon. Their training included several different flavors of interval training, and seven of the eight heart patients completed the race.

Still, even in the late 1990s, many scientists were alarmed by the studies Katharina Meyer was conducting on cardiac rehab patients.

"When I first saw Meyer's work, I shuddered," recalls Carl Foster, the University of Wisconsin physiologist and former ACSM president, who was running the cardiac rehab program at Milwaukee Heart Institute at the time. Foster introduced himself to the German scientist after a seminar she gave at a conference both attended. "Nice talk, Dr. Meyer," Foster said. "How many people have you killed?"

In fact, Foster would come around to introducing interval training to patients in cardiac rehab. One of the key moments in his conversion happened after he wonderingly mentioned Meyer's results to a nurse who had years of experience working with heart patients. "Oh, Carl. You're so stupid," the nurse said, and patted the physiologist on the cheek. Then she took him to a hospital window that overlooked the parking lot, where a rehab patient was making his way to his car. "Watch how this guy walks, Carl," she said, and Foster noticed that the patient could walk only for a short distance before getting winded and having to rest. "They do interval training themselves—because they have to," the nurse told Foster, who has since integrated intervals into the cardiac rehab programs he's managed.

Meyer's work established something counterintuitive about interval training's relationship to the heart. The short, intense spurts of interval training placed a great deal of stress on the working skeletal muscles, like the thighs that power an exercise bike. But because the activity happened for only a short time, the heart muscle didn't have to work as hard as it did during the steady-state exercise typically being prescribed to heart patients. In one study, Meyer compared three different interval-training protocols with the sort of moderate, continuous exercise more commonly prescribed to patients with heart failure. She found that the interval training protocols stressed the heart by an amount between 12 and 17 percent *less* than did the traditional aerobic workout. Meyer's research suggested that so long as a doctor had concluded the patients' cardiovascular disease was stable, the new technique actually was significantly safer for them than steady-state aerobic training.

Scientists working out of St. Olav's University Hospital in Trondheim, Norway, extended Meyer's research the following decade. In 2004 they conducted a comparative study that gathered a small group of patients with cardiovascular disease—patients who'd had things like heart attacks, surgery to insert artery-opening stents, or an arterial blockage so severe it required bypass surgery—and asked them to conduct exercise on an inclined treadmill. The volunteers were separated into two groups. Over the course of ten weeks, both groups conducted three training sessions a week, with half conducting regular steady-state exercise, and the other half conducting interval training. The interval-training group exercised for thirty-three-minute sessions, including five minutes of an easy warm-up and then four reps of four-minute sessions separated by three-minute intervals, with a maximum intensity of 85 to 95 percent of their VO_{2peak}. To match the work done by the interval group, the steady-state group conducted forty-one minutes per session at 50 to 60 percent of VO_{2peak}.

After the ten-week intervention, the steady-state patients boosted their fitness by a little—and the interval-training group boosted their fitness by a lot. In fact, the interval-training group more than doubled their improvement in cardiorespiratory fitness compared with that of the steady-state group. "The results of this randomized controlled study demonstrate that high intensity aerobic exercise is superior compared to moderate intensity exercise for increasing [cardiorespiratory fitness] in stable [coronary artery disease] patients," the study authors concluded.

The results were so promising that St. Olav's University Hospital acceded to a bigger study, this one a randomized clinical trial, which was published in the prestigious journal *Circulation* in 2007 by Ulrik Wisløff and his team. It compared subjects who performed intervals at high intensities with those who performed continuous exercise at moderate intensities. The subjects were patients with heart failure, most of whom tended to be excluded from scientific studies like this because it was considered too dangerous to test them. Previous to Wisløff's, most other studies excluded subjects who were older than seventy. In contrast, the *average* age of Wisløff's twenty-seven subjects was seventy-five. They were stable, but pretty out of shape, with one measure of their cardiorespiratory fitness coming in at just 13 milliliters per kilogram per minute. That's incredibly low. To put that figure into context, it allows a maximal effort that's only three times the resting-energy expenditure. The average sedentary person can perform maximally at ten times the resting-energy expenditure, and endurance athletes at more than twenty times that. The group that Wisløff studied would have been completely winded taking a single flight of stairs.

Some of the subjects conducted moderate-intensity training three times a week, which entailed forty-seven minutes of incline walking on a treadmill at 70 to 75 percent of their peak heart rate. The interval group warmed up for ten minutes and then conducted four repetitions of four-minute intervals that were hard enough to get the heart rate up to 90 to 95 percent of peak heart rate, with each rep separated by three-minute rests. Including the cool-down, the interval group exercised for nine minutes less than the continuous group, at thirty-eight minutes a session.

So how did they compare? The interval group's cardiorespiratory fitness, that all-important indicator of mortality risk, increased three times as much as did the continuous group's. The boost was a remarkable 46 percent for the interval group, compared with 14 percent for the continuous group. That's an *enormous* increase in cardiorespiratory fitness for patients in cardiac rehab.

But the remarkable thing about Wisløff's study was the way interval training changed the patients' hearts. It suggested that you could actually reverse some of the damage that had occurred in people with heart failure. The end result was an improved ejection fraction, or the amount of blood that gets pumped from the heart when it contracts. Recall that in heart failure, the heart does not pump blood efficiently. Often the heart muscle gets enlarged, like an overstretched balloon. The goal of an intervention is to return the heart's pumping capacity—restoring the basic function of the organ so that the muscle walls can squeeze tight and effectively pump blood through the rest of the body.

In the interval-training group, the amount of blood the heart was able to pump improved by 17 percent; in the continuous group, there wasn't any improvement. In fact, by numerous indicators, intense exercise conducted in an interval format helped to reverse the damage of heart failure far more effectively than continuous training, according to the findings of Wisløff and his team.

The notion that cardiac rehab should include physical activity is now taken as a given. Heart centers even have their own workout areas, and there are exercise guidelines for cardiac patients. Most of the exercise prescribed to patients is continuous moderate exercise—steady and sustained exertion on an exercise bike or a treadmill. The problem is, many people don't do it. According to Maureen MacDonald, only 25 percent of cardiac patients actually engage in *any* exercise-based rehab program. Even more troubling? Only 6 percent ever complete their programs, according to that 2015 paper by the American

College of Cardiology's Sports Leadership Council. Those who don't engage in the rehab programs cite a lack of time as the main reason. So is there a better way? For example, what about more time-efficient cardiac rehab—with intervals?

MacDonald was the senior author on a study that compared a typical endurance-based cardiac rehab program with an interval-based approach that required a lot less time. Published in 2013, the study involved men and women in their sixties who were overweight and who suffered from cardiovascular disease so serious that they'd experienced a major event, such as bypass surgery or a heart attack. Twice a week, the endurance group pedaled an exercise cycle at moderate intensity, for thirty minutes at a time at the study's beginning and fifty minutes by the study's end. The interval group also conducted their exercise twice a week, except they performed an interval protocol that involved pedaling hard on an exercise bike for one minute and then pedaling at an easy pace for the following minute, repeating this minute-on, minute-off format ten times. By the end, the interval training was taking about half the time of the endurance training, including warm-up and cool-down. And yet, the interval trainers experienced similar benefits, such as improvements in VO_{2max} and arterial health.

Afterward the researchers asked the subjects what they thought of the interval workout. "They called it a 'really good workout,'" MacDonald says. "They said, 'I could do it. It was hard, but I could do it.'"

Then MacDonald laid out an argument popular with physiologists to justify the incorporation of interval training into more cardiac rehab programs. If "lack of time" is the biggest excuse people provide when asked why they don't participate in cardiac rehab programs, she said, perhaps "time-efficient" exercise programs like high-intensity interval training could provide the benefits of exercise to more people.

Stanford's Jeff Christle says that if he had a relative with heart failure, he would suggest that they undergo some sort of cardiopulmonary testing to ensure that intense exercise is safe for them. And then, he says, "I would want them to exercise at the highest intensity they could."

How safe are intervals for cardiac rehab patients? Ulrik Wisløff conducted a multicenter study that examined the potential risk of intense exercise in cardiac rehabilitation patients—a group that would be considered at high risk to begin with. Three centers across Norway enrolled patients into programs that incorporated what I've come to call the Norwegian protocol. The patients walked on treadmills set to an incline, with the speeds set to elicit certain heart-rate targets. For example, during ten minutes of warm-up, the heart rate target amounted to 60 to 70 percent of the peak heart rate. The main phase consisted of four intervals of four-minute-long duration, with the intent to get the heart rate up to at least 85 percent, and no more than 95 percent, of peak heart rate.

After seven years of tracking, from 2004 to 2011, the research team had data on a group of 4,846 patients with an average age of fifty-eight. Before the exercise intervention, all of them had experienced a diverse range of heart problems, including heart attacks and heart failure. All the patients engaged in both high-intensity and moderate-intensity exercise, with HIIT forming about a third of the total number of workouts. The moderate-intensity exercise was the same incline-treadmill walking used for HIIT, conducted at a constant speed that kept heart rate under or equal to 70 percent of the peak.

So how many experienced adverse cardiac effects as a result of the exercise they conducted? Over the course of seven years, the researchers tracked 175,820 exercise sessions of about an hour each. During or immediately after 129,456 hours of *moderate* exercise, one patient had a fatal cardiac arrest. And during 46,364 hours of high-intensity interval training, two patients experienced nonfatal cardiac arrests. The essential conclusion from the study was that the risk of a cardiovascular event is low after both high-intensity exercise and moderate-intensity exercise in a supervised cardiac-rehabilitation setting.

Interval Training Fights Diabetes

The less active you are, and the more extra weight you're carrying, the less your body is able to control your blood sugar levels. Eventually things can get so bad that the lack of activity and the extra weight can push sedentary and overweight people into a condition called insulin resistance, or prediabetes, which, if left unchecked, can then lead to full-blown diabetes.

Diabetes really impacts a person's quality of life. Wounds don't heal as quickly, and there's fatigue, a feeling of perpetual thirst, frequent urination, tingling in the extremities, and blurry vision. Eventually, you can go blind, and in extreme cases limbs need to be amputated. Diabetes also puts you at greater risk for all sorts of unpleasant, dangerous conditions. For example, those with type 2 diabetes are 50 percent more likely to have a stroke than those in the general population. Also, managing diabetes can be tricky. There are drugs that reduce blood sugar levels, but they come with side effects. In extreme cases, those with type 2 diabetes will even have to give themselves regular injections of insulin.

This is particularly a big problem in the United States. An estimated 60 to 75 million Americans are insulin-resistant and almost 30 million people have diabetes, with 90 to 95 percent of those having type 2 diabetes. And the vast majority of these cases are related to lifestyle factors, such as physical inactivity and a poor diet.

Two great ways to help manage insulin resistance and mitigate the effects of type 2 diabetes are exercise and weight loss. Current physical activity guidelines from the American Diabetes Association recommend that people with diabetes get at least 150 minutes a week of moderate-intensity exercise. The problem is that many people feel they don't have the time to fit in this required amount of exercise. So do we have any indication there might be another way? A more time-effective method to improve blood sugar control?

Yes—and I bet you can guess what it is.

In 2010, my lab conducted a study on the effects of interval training in people with type 2 diabetes. One of the main problems for these individuals is that their muscles don't properly suck up the sugars from the blood. We knew that interval training was particularly effective at rapidly improving the muscles' ability to take up sugar. "We just thought, if your muscles aren't good at taking up glucose," recalls my then–graduate student Jon Little, now a professor of kinesiology at the University of British Columbia's Okanagan campus, "then maybe through interval training we could engage more muscle fibers, which would use up our carbohydrate stores and then suck more glucose out of the blood."

Our original study on type 2 diabetics was a small one: eight people. Their average age was sixty-three, and the average body-mass index was 32, or well into the obese category. Over two weeks we asked them to conduct a modified interval-training protocol we'd developed to try to make the benefits of the training more accessible and less scary to sedentary people. Similar to Katharina Meyer's training program for patients with heart failure, our protocol involved hard (but not all-out) efforts that lasted one minute. Our subjects performed a total of ten repeats on an exercise bike, with one minute of rest in between, for a total of twenty minutes, not including warm-up or cool-down, a protocol that we subsequently dubbed the Ten by One. How hard are we talking? On average the resistance on the exercise bike elicited about 85 percent of the subject's maximal heart rate during the intervals. By the end of the workouts, our subjects were sweating and out of breath—but they didn't find the activity unbearable. In fact, on a 10-point rating scale, with 10 being the most intense effort possible, the subjects on average rated the first interval about a 5 and the last interval about an 8. We had them conduct this protocol six times over two weeks. Then we compared their ability to manage their blood sugar levels before and after the two weeks, to see whether the exercise had changed anything.

The training period was short. I mean, just six sessions over fourteen days. To quantify this in a different way, we were asking our subjects to conduct just an hour's worth of hard exercise throughout the entire experiment. Regardless of it being such a small total dose of exercise, some people wondered whether the older diabetics would even be able to perform the protocol. Would the experiments be a bust altogether?

First of all, they needn't have worried about our subjects handling the training. These people were champs. People have all sorts of misconceptions about the obese and the sedentary. Jon Little has more experience than almost anyone on earth studying the effects of interval training on people with type 2 diabetes. He's noticed a couple of things. One, he tends to be impressed with how hard they work. Two, he's found that many obese and sedentary people tend to find the Ten by One protocol easier than going out and walking for thirty or so minutes. Possibly because it doesn't take much effort to get these people up to 85 or 90 percent of their maximum heart rate.

Back in 2010, once our subjects had gone through their two weeks of training, we inserted these little glucose-monitoring devices under the skin in their abdomens. The devices tracked glucose levels through the course of a full day, taking measurements every five minutes. They allowed us to analyze how well the subject's bodies managed blood sugar levels over a twenty-four-hour period, including when they ate standardized meals.

Little was the researcher who conducted the analysis on the first subject. He was in the lab, saw the curve on the computer screen, and was so excited by what he'd seen that he sprinted upstairs to my office. "Holy smokes!" he exclaimed. "Look how much better we made her glucose after just two weeks of training!"

The results were much better than any of us anticipated. That hour's worth of exercise turned out to have some remarkable effects. The average blood glucose concentration decreased by 13 percent.

If sustained over time, this kind of decrease could improve the body's ability to manage blood sugar enough to actually reverse the onset of type 2 diabetes.

That's how remarkable these results were.

Type 2 diabetes takes a long time to develop—between ten and fifteen years of consistently high levels of insulin resistance, or prediabetes. That means more than a decade of sedentary living and poor diet.

Many people find the prospect of 150 minutes a week of moderate-intensity exercise overwhelming. And that's especially true of people who are so sedentary that they have insulin resistance. They know they should exercise, they might even *want* to do it, but they just don't get around to it. Instead they resort to controlling their diabetes with drugs and, in extreme cases, insulin injections.

Our 2011 study and others since have shown that high-intensity interval training is a potent way to help people manage their blood sugar and decrease symptoms of insulin resistance and type 2 diabetes. It could be enough to nudge some people off the road they're on—a road that leads to diabetes. "When you try to statistically tease out the causes [of diabetes], 70 percent of diabetes cases in the United States could be prevented through lifestyle [changes]," says Little. "It's as simple as eating real foods and exercising."

How Do Intervals Fit into the Guidelines?

How did we arrive at this guideline that we need 150 minutes of moderate-intensity exercise a week to promote and maintain health? The fact is, continuous moderate exercise—the long hours of walking, jogging, swimming, cycling, and their variants—get the emphasis from the public-health guidelines mainly because this type of physical activity has been most studied in the academic literature. The vast majority of this research is epidemiological in nature, where large groups of people are followed over time. In fact, no randomized clinical trials have directly tested whether physical activity extends your life or prevents you from getting cardiovascular disease. We have really good data to show that if you do a lot of long, slow steady-state exercise, you have a lower risk of dying and a lower risk of getting type 2 diabetes and cardiovascular disease, as well as some types of cancer. There are lots of studies about the long-and-slow method because that's how most people get their physical activity.

We have much less data on the benefits of interval training. In fact, think of interval training as a recently developed drug—after all, its powerful and acute effects rival those of any drugs. Drugs must undergo a long and arduous process before government regulators approve them for medical use. The process typically begins with basic research, progresses through small-animal studies, moves on to small-group human studies, and culminates in long-term randomized clinical trials.

Dozens of such studies have found that moderate-intensity continuous training improves human health over the long term. Interval training is so cutting-edge that it's still in the early stages of testing. Ulrik Wisløff is the senior author on the first randomized and controlled clinical trial on high-intensity interval training and human health. In progress since 2012 in Trondheim, Norway, the Generation 100 study hopes to wrap up in June 2018, having included more than 1,500 people in its exercise interventions over five years.

A professor of epidemiology at Harvard University, I-Min Lee is a dual MD and PhD who studies human health on a population level. She specializes in the way exercise affects the health of populations and helped create the US public-health guidelines on exercise.

She says the Generation 100 study is the sort of evidence

that public health organizations want to see before they incorporate specific language about high-intensity and sprint interval training into their exercise guidelines.

That said, Lee believes the guidelines are robust enough, and general enough, to encompass interval training. (I'll delve deeper into the various flavors of interval training, including examples of different workouts, in chapters six and seven.) "The guidelines don't specifically mention these protocols, but sprint interval and high-intensity interval training do fall under the broad umbrella of vigorous exercise. . . . Individuals should fulfill [the guidelines] in whatever way they choose."

So if interval training fits within the exercise guidelines, how do you "count" the workouts against the guidelines' weekly totals? Recall that the American College of Sports Medicine calls for 150 minutes of moderate exercise per week or 75 minutes per week of vigorous exercise. And let's say the interval workout we're discussing is the Ten by One protocol employed in the studies I mentioned earlier on people with cardiovascular disease and diabetes. The protocol involves ten hard efforts of sixty seconds each, and lasts about twenty-five minutes when you factor in the warm-up, cool-down, and recovery periods between intervals. When determining how to "count" that against the guidelines, do you include just the ten minutes of hard effort? Or the workout's total duration of twenty-five minutes—including the recovery periods?

As she notes, Lee believes high-intensity interval training belongs in the category of "vigorous" exercise. As to whether you count the sprint duration or the whole workout's duration,

Lee answers by bringing up other activities that qualify as vigorous, like singles tennis, basketball, and ice hockey. In those cases, the guidelines would allow you to count the total time that you were playing the sport—even though each game includes many moments of inactivity, such as the pauses between tennis serves or the time that basketball players spend standing around as someone takes a free throw.

Applying that same logic to high-intensity interval training, Lee says that you count against the guidelines the time required to conduct the entire workout. So if the whole Ten by One workout requires twenty-five minutes of your time, you would have to repeat the workout three times a week to fulfill the spirit of the guidelines, according to Lee. The point is, you don't just count the time spent doing the hard sprints. The recovery periods count as well.

By now, whether we're talking about varying your speed on a walk or all-out twenty-second sprints, I hope I've convinced you there's likely an interval-training method that's appropriate for you.

CHAPTER FIVE

High-Intensity Engagement

It's midday on a Saturday, a fitness studio in midtown Manhattan. The clients arrange themselves into two lines and then, at the trainer's signal, explode forward and back across the padded turf floor. There's a bear crawl, a crab walk, frog hops, and high steps—all performed as fast as possible. At various points the trainer tells the clients to perform sets of burpees, push-ups, crunches, and squats. Resting after the first go-round, the clients have already started to breathe hard. By the third, their athletic clothing is dark with sweat. Everyone here looks fit, but this workout is so tough it's taxing even the best of them.

The session at Manhattan's Tone House gym, renowned to

be one of the toughest workouts the city has to offer, is based on the principles of interval training. It's great—for those who can do it. Unfortunately, associating the training protocol with such tough workouts also comes with a major drawback: It prevents a lot of people from accessing interval training's benefits.

Mention the term "interval training" and many people envision football players grunting through all-out sprints on some gridiron grass, or ripped fitness enthusiasts leaping into infinite sets of box jumps. They think it's not for them. I can understand having that sort of reaction. If I didn't know anything about the exercise method, I'd probably feel that way, too.

But the stereotype is wrong.

The fact is, interval training is something that has a lot of variation in it. Sure, it *can* be hard-core. But it doesn't *have* to be. Retirees and heart patients can do one flavor while elite athletes can perform another. The basic idea of varying the intensity of a workout to make it more effective can be used to create programs that increase fitness for people of all ages and lifestyles.

Anyone in the exercise trade will tell you that one of the hardest things about getting fit is that, sometimes, for people who are just starting out, exercise *hurts.* You spend your whole life hearing about stuff like the "runner's high" and how good people feel when they exercise. Then you do it, and darn it if all that doesn't seem like a lie. Exercise isn't this friend who makes you feel good. Not at first. Exercise makes you feel like hell. Exercise is a *jerk.*

That period goes away—and if you're using interval-training techniques, it can go away more quickly. In fact, two former graduate students of our program, psychologist Mary Jung and physiologist Jon Little (from the previous chapter), were involved with a study that shows this. Published in 2012 and led by researchers out of Queen's University (Kingston, Ontario), the study asked young women to conduct a brief fitness protocol four days a week for a month. Each day, the young women ran through a Tabata-style format of twenty seconds on, ten seconds off, repeated eight times for a total workout duration of four minutes. The "on" portion saw the women conducting one of the following bodyweight exercises: burpees, jumping jacks, mountain climbers, or squat thrusts. It was a bit like the Tone House workout actually—except a lot shorter. The whole workout required about sixteen minutes a week. The women improved their cardiovascular fitness to the same extent as another group that performed two hours a week of moderate-intensity continuous cycling. And the interval trainers became stronger, too—improving their performance on strength tests like push-ups and leg extensions, an added benefit that did not occur in the other group.

More to the point, though, after experiencing the remarkable benefits of interval training, the young women decided they liked it more than they'd expected to. In questionnaires, the young women said they would be more likely to exercise using interval-based workouts in the future.

The study's results reflect the way that the first, painful phase of exercise is temporary. When you use interval-training techniques, it lasts just a couple of weeks. In fact, interval training makes that awkward period that coincides with exercise's beginning as short as possible. It gets you fit *faster* so that you can begin enjoying exercise sooner.

We're not even talking about the most intense flavor of interval training here. Pretty much any type of interval training, which sees you varying the intensity of your effort throughout the workout, is going to benefit you more, and more quickly, than a comparable period of continuous training. You go out for a walk, and you vary the intensity of your effort? That's interval training, just as much as is the variety that sees the human equivalent of ultrafit rabbits and gazelles powering through spin classes and the sort of circuit-training workouts that happen at Tone House in New York.

The Weird and Wonderful Thing About Exercise

Like a lot of people who have researched the science of fitness, I regard exercise as having near-magical properties. We spend billions on research to create pills that improve health. Most of these pills target only one aspect of health, and many have unwanted side effects. Meanwhile, the most powerful intervention possible—exercise—goes comparatively underutilized. I mentioned earlier in the book that only about 15 to 20 percent of the people in the United States, Canada, and the United

Kingdom actually get the 150 minutes of moderate exercise a week that the guidelines suggest. That means anywhere from 80 to 85 percent don't get the recommended amount of exercise, even though it benefits your physical and mental health in all sorts of ways, from helping you live longer to enabling you to live a happier life. Exercise is the best way to push back the effects of aging. It's the best way to hedge against a decrepit last few years of your life. Want to live an active life until you're ninety? The best way to promote that is by exercising regularly. And yet so few of us do it!

That's a problem. The human lifespan in the developed world has doubled in the last two centuries, and yet human "healthspan," or the proportion of our lifespan that we are active and healthy, has not kept up, according to a 2015 *Journal of Physiology* paper lead-authored by the British physiologist Ross Pollock. The consequence? "The years spent with poor health and disabilities in old age are increasing," Pollock and his coauthors wrote.

Alan Batterham is an exercise scientist at Teesside University in Middlesbrough in the United Kingdom. He says there's a perversity to the human relationship with exercise. Humans are lucky that most of the things that are good for us also happen to be programmed into our gray matter. If we need water, for example, the body sends a thirst signal to the brain, and we drink. But no such need-drive relationship exists with exercise, which may be the only activity our bodies require for optimal health that we're not biologically preprogrammed to carry out. "There is apparently no innate drive to be substantially physically active," Batterham observed.

That's probably because, in ages past, exercise was something we got enough of purely by doing the things we had to do to survive. In fact, as Harvard University paleoanthropologist Daniel Lieberman points out in a 2015 paper, humans evolved a drive to *avoid* physical activity—to conserve energy when possible. We did that, according to Lieberman, because thousands of years ago, food was scarce, as was the energy that we derived from food. If you were a hunter, then you needed to conserve those calories to use as you sought your prey. If you were a gatherer, you needed those calories for fuel as you wandered about your habitat, searching out grubs, berries, and other sustenance.

Things are different today. Thanks to the industrial revolution, urbanization, the automobile, and the development of computer technology, most people in the developed world are able to procure their daily sustenance without expending much physical effort. Many times, all that procuring food requires is a walk from the couch to the kitchen. Now, that biological imperative toward rest has created an exercise paradox—that "people tend to avoid exercise despite its benefits," Lieberman says. Somewhere, we're aware that our bodies would work better and we'd feel happier if we increased our activity levels. Yet the pace of life is such that most of us feel we just don't have the opportunity.

A Rift in Academia

Physiologists in general tend to be pretty excited about interval training for all the reasons stated above. But the debate among psychologists is more fervent. Opposition to the technique is exemplified by a study written by a psychologist from Curtin University in Perth, Australia. Her name is Sarah Hardcastle, and she specializes in exercise.

In a 2014 article published in *Frontiers of Psychology*, Hardcastle argues that sprint interval training, the most time-effective and intense flavor, is too hard for most people to do. "Proponents of SIT [sprint interval training] have focused almost exclusively on physiological adaptations," Hardcastle and her coauthors note. Then they ask whether a sedentary population "will feel physically capable and sufficiently motivated to take up and maintain a regime of highly intense exercise." Sprint interval training, says Hardcastle, "is likely to be considered too arduous and may evoke anticipated perceived incompetence, lower self-esteem and potential failure." The article's title? "Why Sprint Interval Training Is Inappropriate for a Largely Sedentary Population."

Sure, interval training is really effective, Hardcastle says, but for most people it's too hard—physically *and* psychologically. She goes even farther than that, however, arguing that "high intensity exercise is likely to evoke negative affect which may lead to subsequent avoidance of further exercise." In other words, sedentary people will find sprint interval training workouts so unpleasant that the intense physical activity will discourage them from doing other types of exercise in future. That sprint interval training could put them off exercise altogether.

There are people out there who subscribe to Hardcastle's argument. Some highly respected exercise psychologists have completely dismissed high-intensity interval training as a mechanism that can affect public health. They just don't think people are going to do it. The general belief in exercise programming in the United States is that people would prefer to exercise at a low intensity over a long time rather than at a higher intensity for a short time.

"The problem of sedentary living is large, *particularly* in the US, and it's such that people are *scared* of suggesting interval training," says Stanford University's Jeff Christle. "The fitness industry has gotten a hold of this thing, and in the US they see that as a big deterrent. They're just concerned about 'Let's get people *moving*. Why make it more complicated than that?'"

In June 2015 the debate flared up again at the annual meeting of the International Society for Behavioral Nutrition and Physical Activity, which that year occurred in Edinburgh, Scotland. Before a crowd of approximately four hundred, in the Pentland Auditorium of the Edinburgh International Conference Centre, psychologist Stuart Biddle of Australia's Victoria University argued that interval training's potency was "largely pointless"—because few people would ever do it. The problem isn't that people don't have the *time* for exercise, Biddle argued. Rather, they're just not making it enough of a priority. To encourage people to make exercise a priority, Biddle said, according to a transcript of the debate published in an academic journal, "we need to boost people's positive feelings about exercise. Making it harder and more painful"—as he claims interval training does—"is unlikely to do this."

As it turns out, however, many people find interval training *more* enjoyable than conventional exercise.

A Short Course on the Psychology of Exercise

I've been interval training for a decade now, and I've loved it ever since I started. So when I first heard the psychologists' objections that people wouldn't do it, I shook my head. To me, intervals are a completely different animal from continuous vigorous exercise. Yes, it's hard to maintain a strenuous pace for a prolonged period of time. But intervals are different. You get breaks from the toughest part of the exercise, which allows you to recover and makes the hard work seem easier. Intervals keep you engaged, which makes the time feel like it's going even faster than it is.

I couldn't refute the objections of such figures as Hardcastle and Biddle with my *own* experience, however. So I dug into the research on the psychology of interval training. One of the most interesting studies I found was conducted by a team out of Liverpool John Moores University in the United Kingdom. They published a study in 2011 that examined whether people preferred interval or continuous running. The neat thing about this study, which was conducted with recreationally active men, was that the researchers tried to match the protocols as much as possible—that is, the average exercise duration, intensity, and distance run were the same. The continuous workouts required the men to run at 70 percent of VO_{2max} for fifty minutes—a good, moderate pace. The interval version required the men to run six three-minute reps at 90 percent of VO_{2max} interspersed with three-minute breaks. The workouts also matched in terms of their average heart rate.

Afterward, the men were asked to rate which workout they enjoyed more using something called the Physical Activity Enjoyment Scale, and to rate the extent they exerted themselves during the workout. The men believed they exerted themselves more during the interval workout—and yet, *they rated the intervals as much more enjoyable than the continuous running*. The average enjoyment score for the interval running was 88 while the continuous exercise was down around 61. So here was a study that suggested the old supposition—that higher exertion equals less enjoyment, as well as a lower likelihood to continue performing the exercise—didn't apply to interval training.

Ah, the naysayers said. That British study didn't ask the subjects to rate their enjoyment until several minutes *after* the exercise had ended. To figure out whether people will actually do the exercise, in the wild, you have to ask how people feel *while they're doing the exercise*. (The other problem with the

British study, critics said, was that its subjects were fit young men who might be more accustomed to exercise and not experience as much discomfort.)

One Psychologist's Take

As an exercise psychologist, Mary Jung was in many ways the perfect person to attack the problem of interval training's likability. I got to know Jung when she was a master's student in our kinesiology graduate program at McMaster. At the time, she was working in the next lab over, studying how to best encourage dieters to stick with their eating plans. Even through the lab's steel doors and the cinder-block walls, Jung recalls how she would hear us clapping and cheering on our experiment subjects through the course of their training interventions. (We were working on one of our all-out sprint studies at the time.)

So she came over to check us out. She got one look at a test subject working as hard as she could on an exercise bike, with the music cranked and everyone gathered around to watch. Jung didn't have any idea what we were doing, she would recall later, but she wanted to be a part of it. And as time went on, at my lab and others, Jung applied a psychological viewpoint to the study of interval training. Her work examines how much subjects enjoyed interval training—and whether they would continue the training method in future.

Jung's interest in the psychology of exercise far predates her time as a graduate student. In high school, she started teaching fitness classes and continued for sixteen years: spin classes, aerobics, step classes, boot camp workouts, kickboxing—you name it, she taught it. She also has more than a decade's experience as a personal trainer. You know, in addition to her PhD in exercise psychology. She currently runs the Health and Exercise Psychology Lab at the University of British Columbia's Okanagan campus, where her research interest resides at the intersection of exercise and psychology. For example, she investigates how it is that some people develop great exercise habits while others resent having to walk to the corner store to buy snack food. Why do some people stick with workouts while others can never keep to their routine for longer than a week or two?

Back in 2013, Jung encountered critics claiming that the various iterations of interval training were too difficult to conduct, and she got a little frustrated. She'd witnessed firsthand how people enjoyed the variety and potency of high-intensity interval training. Not only did Jung know that other people liked HIIT workouts; she herself has participated in and enjoyed HIIT. "I like how quickly it works," she says. "And I can see how others respond to that, too. Using HIIT, they notice the changes brought on by exercise a lot faster and a lot earlier once they start working out."

Jung became as hopeful about interval training's potential to affect public health as I was. She, too, regards it as a way to bring the benefits of exercise to more people—because interval training's reduced time commitment will eliminate the barriers that prevent people from exercising. And she's witnessed firsthand people's sense of pride and accomplishment after they've done a particularly rigorous interval workout.

Jung believes that interval training is peculiarly suited to healthy but sedentary people just beginning exercise regimens. "What is considered high-intensity for an individual who has not exercised in years looks drastically different than conjured-up notions of all-out exercise performed by athletes and in boot camp classes." Someone who is completely out of shape could gain the extra-potent benefits of interval training, Jung believes, purely by walking up a slight hill, turning around to go back down, and repeating that a handful of times.

"There is a perception among people, particularly among those who don't exercise regularly, that interval training is not for them," Jung says. "But many people don't understand what interval training is. They have visions of these Army-style boot camp workouts. They ask, isn't HIIT only for elite athletes? And that perception may be keeping the benefits of interval training away from the people who could most use the benefits."

The Comparison Study

The next step for Jung was to conduct a scientific examination that compared how beginner-level exercisers felt during HIIT and during moderate continuous exercise. It was one of the first times an investigator had looked at this particular facet of HIIT. Jung aimed to make the study as realistic and practical as possible. Through posters at the Okanagan campus, she and her fellow researchers recruited forty-four young people who were not exactly sedentary but certainly not athletes either—people who worked out less than twice a week. Then she asked them to conduct three different workouts.

The first was a moderate-intensity continuous bout on an exercise bike that lasted forty minutes, saw them working out at 40 percent of their maximum power output, and raised the subjects' average heart rate to 69 percent of their peak. A second was a vigorous intensity continuous workout that saw them working out for twenty minutes at 80 percent of their maximum power output, which raised their average heart rate to about 89 percent of their peak. And the third was the interval workout. For this study Jung used the Ten by One protocol that we had used for less athletic subjects, such as sedentary people and those with diabetes.

At various points during and after the three workouts, the subjects were asked to rate how they were feeling. Then, twenty minutes after they finished, the researchers asked them how likely they would be to engage again in the type of workout they'd just completed.

According to their own ratings on an 11-point scale, the subjects felt good during the continuous moderate exercise, just OK during HIIT, and pretty bad during the continuous vigorous exercise. But one of the most interesting things about the study was the improvement in the subjects' self-rating that occurred twenty minutes after the HIIT workout. Jung calls this the "positive rebound effect." In other words, twenty minutes after the workout was over, all three groups felt much better than they had during the workouts—but the moderate and interval groups felt the best.

Next, asked to rate their likelihood to engage in any of the three workouts in the future, the subjects said they were equally likely to conduct continuous-moderate and interval-training workouts, while they were less likely to conduct continuous vigorous exercise.

Jung and her coauthors then surveyed the group about which activity they had most enjoyed. HIIT was ranked as slightly more enjoyable than the moderate exercise, which in turn was rated as more enjoyable than the vigorous exercise.

Finally, when asked to name an exercise choice as their favorite, the subjects said they far preferred HIIT. Twenty-four of them identified intervals as their favorite exercise, while thirteen preferred continuous moderate exercise, and four preferred continuous vigorous exercise. (Presumably, the remaining three indicated no preference.)

So the subjects felt just OK while they were *doing* the HIIT workout—but then overwhelmingly chose it as their favorite? Does that make sense?

Mary Jung believes it does. The interval approach, she says, "breaks down the exercise sessions into short, surmountable bursts, potentially allowing for multiple successful experiences." In other words, when the subjects completed a sprint, they felt as though they'd *accomplished* something. And that sense of accomplishment offset the discomfort of the more strenuous interval. The workout made them feel like *exercisers*—a feeling they'd never had before. In addition to that, factor in the time savings. Taken together, that's all enough to justify why the subjects favored interval training so much more than the other two workout forms, according to Jung.

Jung believes—and I agree with her—that her study shows that people are more sophisticated than some exercise psychologists believe.

The old thinking, which suggests that most people aren't likely to perform difficult exercise, doesn't seem to apply to interval training. Not that intervals *have* to be all that difficult. The most strenuous versions of interval training require you to push it. When you're starting out, that can hurt. And what Jung's study shows is that people take in the time-efficiency benefits of HIIT, factor in the boost of confidence that happens after they complete an HIIT workout, and then also consider the fact that interval workouts tend to be less boring than conventional endurance training—and they weigh all that against the temporary discomfort they experience while doing a short interval. And they decide that HIIT is worth it.

Jung has also demonstrated that people prefer HIIT for yet another reason. She's spent the past several years following sedentary and diabetic individuals through interval-training programs. While lots of people are doing that, it's the way that Jung designs these workouts that makes them significant. In the usual template, she asks a certain subgroup of the population to perform a fitness training program—some pursue moderate continuous exercise; others pursue interval training.

Then Jung sits back and waits. Weeks or months can go by, depending on the study. And once some time has passed, she checks in with the group to see who still continues to exercise. The point is to conclude which sort of workout sets up people to exercise regularly in the future. To increase that proportion of people who *do* get the activity the guidelines suggest well beyond 15 to 20 percent.

For example, in 2015, in the *Journal of Diabetes Research*, Jung and her coauthors published a study that followed twenty-six adults through one of two twelve-day exercise programs, one featuring only intervals, the other featuring moderate-intensity continuous exercise. These were middle-age adults who were pretty out of shape and qualified as prediabetic. The moderate group could manage only twenty minutes of continuous exercise at a time and gradually worked their way up to fifty minutes. The interval group could manage only four one-minute intervals separated by a minute's rest, and over the course of the twelve days they worked themselves up to our standard Ten by One workout. Then Jung and her group left the subjects alone for a month.

When they checked in with the subjects again after the month had passed, they found that the interval-training group was a lot more likely to perform vigorous exercise on their own than the group who had conducted moderate-intensity exercise. In fact, a month after the training program ended, the interval group was *doubling* the amount of vigorous exercise they'd conducted previously. A similar study, which was not yet published while I was researching this book, reveals analogous results *after six months*.

So what's going on? Jung believes that interval training is more effective than moderate-intensity continuous exercise *at selling people on the benefits of exercise*. She says, "From a psychological perspective? For people who have spent most of their lives avoiding exercise because they think it isn't for them? They go through an interval-training program, and they get this massive boost in confidence. It's like, 'Oh my gosh, I *did* it.' They think to themselves, 'I may not be an athlete, but I really worked it.' They get a sense of accomplishment. You don't get that same rush from walking around the block at a moderate pace. And from an exercise-adherence perspective? That confidence is really important."

There's also the fact that physical changes happen sooner in people who do interval-training workouts than in those pursuing moderate-intensity continuous training. Stairs become a lot easier to climb. In general, they just feel better, *faster*, with interval training. "They're happier with the results," Jung says. "That elicits a good vibe, good feelings, pride. When you feel that, you kind of want to feel that again. And you talk about it at parties—'Hey, I'm an exerciser now.'"

A Few Words for Those Just Starting Out

First, remember that interval training has a lot of different flavors. You can enjoy its benefits simply by varying the intensity of an aerobic workout. You push hard, you back off, you push hard again. Congratulations. You've just performed the same sort of workout that's been used by Olympians and world-record holders.

I believe interval training is perfectly appropriate for beginning exercisers to do. And I believe that because of the way that sedentary people tend to work out when they first start exercising. Allow me to explain.

Mary Jung and Jeff Christle both have led studies that have asked sedentary populations to conduct exercise—the sort of subjects who some believe should run away screaming from interval training. Jung has worked with diabetics as well as obese and sedentary people on the verge of metabolic syndrome; Jeff, with people who have had heart failure. And both Mary and Jeff say the same thing.

People who are deconditioned—completely out of shape, perhaps because of cardiovascular disease or an unhealthy lifestyle—might set *out* to conduct moderate-intensity continuous exercise. But whatever they tend to do, they have to take rest breaks. No matter how moderate the exercise, they can't keep at it for long.

Bear in mind, some of these people get out of breath going up a flight of stairs. They can't walk all the way around a block in one go. But let's say they try. Let's say something has happened to motivate them: A close friend has had a health scare. Or their doctor has asked them to follow the 150-minutes-per-week-of-activity guideline, and so they set out for a walk. And they can't do it. They can barely get two lampposts down the street before they have to take a break. And that turns out to be a big buzzkill. They *berate* themselves because they weren't able to manage the continuous exercise. And they end their attempt at moderate, continuous exercise by feeling bad about themselves.

But what if we took a different approach? What if we told these people, "Hey, listen. Don't beat yourself up because you weren't able to do a certain amount of moderate continuous exercise.

"Instead, *celebrate* yourself, because you know how that bout of walking just elevated your heart rate so much that you got completely out of breath? You just did an interval! And after you take a break, and you walk the two lampposts back to your house? That's another interval. Two intervals—good going! You've just done an interval workout!"

The people who dismiss interval training as too complicated and too arduous lose sight of the fact that most deconditioned people—the people they most want to reach—end up starting out accidentally conducting interval training *anyway*.

They do it naturally. They set out for a walk, get a ways, take a break. Then they go a little farther, take a break. They head back, go a ways, take a break.

Interval training also seems more manageable to these folks. Let's say you set them up on a protocol that has been shown to work great for various populations: the Ten by One, which asks people to work hard for a minute, then back off, and repeat that ten times for a total of twenty minutes per workout. Ask people to do it three times a week for a total duration of an hour? That becomes a lot more manageable—particularly to the sedentary target population that US public health officials most want to reach.

"Using intervals to get people through the door is a nice way to get sedentary people moving because you benefit so quickly," says Jeff Christle. "Intervals three times a week, the Ten by One—I think that [with that] kind of program, within four weeks they'll see really nice improvements. With a regular walking program, the improvements take twice as many weeks. That's one of the big advantages [of interval training]."

In the next chapter, I map out a series of workouts for everyone from older people who have never exercised to high-performance athletes looking to gain an edge in their next race. But before we get to that, I want to finish this chapter with some tips. The hardest thing about interval training is starting out. Particularly after you've just begun an interval training protocol, the more-intense sprint portion of the workouts can be tough. But here are two tips to make them more enjoyable—or at least, hurt less.

First, music helps. A 2015 McMaster paper I coauthored analyzed how distraction affected sprint workouts. We found that listening to the motivational music of their choice helped our subjects pedal harder and faster than not listening to any music at all. The idea is to get amped up, to psych yourself into a state that makes it easy to reach the high intensity. Some people use dance music; others, gangsta rap or punk rock. For me, old-school Van Halen does the trick.

Second, get a little help from your friends. Whether it's a

buddy shouting, "Go! Go! Go!" or a coach applauding your effort, encouragement helps as much during intervals as it does for anything else. At least, that's the take of a 2013 study out of Western University that examined how verbal encouragement affected enjoyment of an interval workout. When compared with groups who received little or no encouragement, those who got positive feedback during their sprints ended up enjoying their workouts more. After all, some workouts can be tough—so take all the help you can get.

Now, I'd like to present Mary Jung's five tips for starting and sticking with an exercise program.

1. Boost Your Confidence

Developing the confidence to begin exercise can be a chicken-or-egg situation, Jung says. To start out, you have to have the confidence that you can do it. But how do you develop that confidence if you've never done the exercise in the first place? To solve the problem, Jung turns to the work of renowned Stanford psychologist Albert Bandura, who determined some of the most effective ways to develop a belief in oneself. Say you want to start a minute-on, minute-off workout but you're not sure whether you can push yourself for a minute. Think about other times you've broken a sweat; maybe you recently had to run to catch a flight in an airport. Apply that feat to this new situation. "I sprinted to the gate and caught the plane to La Guardia," you might say. "And if I can do *that*, surely I can do *this*."

Another way to develop confidence is to look at the stories of other people who have been through what you're about to go through. Throughout this book we've discussed how cardiac rehab patients and those with type 2 diabetes have started and stuck with interval-training workouts. If they can do it, you can, too.

2. Start Early in the Day

When you're first starting out, Jung suggests scheduling your exercise for the morning, when your willpower is highest. Here, Jung draws on the thinking of the psychologist Roy Baumeister at Florida State University, who oversaw pioneering studies on human willpower. Baumeister established that willpower ebbs and flows throughout the day. It's lower when we're tired or hungry—and taxing our willpower in one way means we have less willpower to exert later for something else. Our willpower stores tend to be highest when we've just woken up after a good night's sleep. So make it easy on yourself, and exercise in the morning.

3. Be Kind to Yourself

We tend to be our own worst critics. The problem is, self-loathing leads to poor decision making. "You forgot to launder your workout clothes again, idiot," a person might think. "Why bother to try to do good things—you're only going to screw them up." Rather than flinging insults, Jung suggests cutting yourself some slack. Acknowledge that setbacks happen. You'll have to skip a workout now and then. Everyone does. The key is to maintain the confidence that you *will* make it to the next workout—and being hard on yourself is no way to maintain confidence.

4. Avoid Comparisons

We're social animals; it's profoundly human to compare ourselves with others. Many of the reasons we begin working out reflect this human tendency. Plenty of people join gyms so that during spring break they'll look better than their friends in a bathing suit. But such motivation tends to be short-lived. It may not make you feel all that good about yourself, either, even if you do meet your goal. The idea here is to set a lifelong exercise habit that boosts your self-esteem. So when you're making your goals, focus on things that don't require you to step on the heads of others as you hoist yourself up. "I'm working out so that I'll have the energy to play with my grandkids" is a lot better reason than "I'm working out so that I can fit into smaller pants than Cecilia can."

5. Reward Yourself

Once you've met some achievable goals, reward yourself. For example, one of Jung's favorite rewards is a glass of wine. It's small and affordable yet feels indulgent. Another option may be time to oneself, away from the kids. Or perhaps a hot bath?

Let's say you're trying to work out three times a week. The evening of the first week you manage that, Jung suggests setting aside some time to indulge in a reward, whatever it is. Rewarding yourself soon after you achieve your goal consolidates the goal-reward relationship and motivates you to achieve other goals in the future.

CHAPTER SIX

Fun and Fast: Eight Basic Workouts

I DESIGN MY OWN WORKOUTS ALL THE TIME. I GO DOWN into my basement, hop on the exercise bike, turn on some sports on the television, and get to it. Sometimes I'll perform various bodyweight exercises between sprints. Push-ups, pull-ups, burpees—you name it. Each of these amounts to strength-building resistance training that will help me stave off the effects of aging. Creating workouts is one of the fun things about exercise. It keeps things fresh.

This chapter is all about providing people with the tools to create their own high-intensity interval training workouts. The difference between these workouts and those in most other exercise books is that the benefits of these programs have been scientifically proven. Below, I include eight workouts that have been featured in peer-reviewed academic studies—protocols used on everyone from cardiac rehabilitation patients to high-caliber athletes. I describe the workouts' scientifically assessed benefits and finish with some tips to help you design your own. But first, a few words about how we quantify "effort."

Fast and Furious: Assessing Intensity

Envision a typical spin class. A trainer straddles a bike that faces a dozen clients on exercise bikes. Shouting loud enough to be heard over a pounding techno beat, she barks into her headset: "Give me 90 percent in three . . . two . . . one . . . And *go*!" All around the room, veterans and first-timers alike mash their pedals. But chances are they're all working at different rates—because they have different perceptions of what a 90 percent effort feels like.

Which brings us to one of the trickiest elements of high-intensity interval training: How do you convey to someone the effort required for a given sprint? Many different ways exist to measure exercise intensity. But most of them have been designed to work with long-duration continuous *aerobic* workouts—and, when applied to sprints, each of them comes with its own idiosyncratic problems.

VO_{2max} and Power

Traditionally, the most precise way to determine workout intensity involves measuring how much oxygen a person uses during exercise. We express that measurement relative to the person's maximal oxygen uptake, or VO_{2max}. So we might say a person is cycling at 75 percent of their VO_{2max}. However, several problems exist when using VO_{2max} to describe the effort expended in intervals. In a sprint, it's possible to run many times faster than the pace needed to elicit your VO_{2max}. And measuring VO_{2max} is unwieldy, requiring a mask over the mouth and nose, and breathing tubes leading to a device that measures the difference in the oxygen content of the air the subject inhales and exhales. It's just not practical outside a lab.

Another scientific measure of exertion is *power*, which is most commonly measured in watts. The same unit is used to describe the brightness of a lightbulb or the energy efficiency of your fridge. A watt is a measure of the rate that work is done. On a bike, healthy but untrained people might reach their VO_{2max} at a power output of 300 watts—but in a sprint, that same person can generate 900 watts for a few seconds, while elite track cyclists can achieve 2,000 watts or more. That's a work rate that could light twenty 100-watt bulbs—enough to illuminate an entire large house. The trouble with that measure? We want each of these workouts to be adaptable to many different forms of exercise. While computers can easily measure watts on exercise bikes, it's hard to determine power output with other activities such as swimming or running.

The Trouble with Heart Rate

Another common way to describe exercise intensity is relative to a person's *maximum heart rate.* (A convenient way to estimate this value is 220 minus your age, although a lot of variation exists between individuals.) "Go for a run at 65 percent of your maximum heart rate," one trainer might say—a rate of exertion that would see a typical forty-year-old aim for an average heart rate of 117 beats per minute. That's a lot easier to measure outside the lab because heart-rate monitors are getting more portable and accurate all the time.

The trouble is, heart rate is also a problematic measure to describe the intensity of maximal and near-maximal sprints. Because heart rate lags effort, especially at the start of intense exercise. "Go as hard as you would to save your child from an oncoming car," I tell people when I want them to give me an all-out sprint. But after a thirty-second all-out sprint, your heart rate might get up to only about 70 percent of its maximum. That's because when you exercise very vigorously, it takes the cardiovascular system a bit to catch up. The only time you reach your maximal heart rate is through hard aerobic exercise sustained over a few minutes, or over the course of repeated sprints with very short recovery periods. So while the percentage of maximum heart rate works well as an indicator of exertion during steady-state aerobic exercise, and it can even be adopted for less-intense sprints, it's a lot more problematic for describing the exertion required to sprint at an all-out pace.

So What's the Answer?

The solution begins with something physiologists call *rating of perceived exertion*, or *RPE*. The psychologist Gunnar A.V. Borg of the University of Stockholm introduced the concept in 1970. Borg tied his rating of perceived exertion to heart rate. The original scale went from 6 to 20—a range that roughly matched the heart rate of an average young person divided by 10; it began at 6 because an average young person's heart rate is around 60, and it went up to 20 because that same young person's heart rate tops out at about 200 beats per minute. Walking, or "light" exercise, would correspond to an RPE of 10 or 11.

A scale from 6 to 20 is a bit unwieldy, so Borg subsequently came up with a simpler RPE range from 1 to 10, with 1 being described as "nothing at all" and 10 as "very, very strong (near max)."

The intensity scale of the workouts described in this book is modeled on Borg's revised RPE scale. Note that, as with Borg's scale, it's possible to go higher than 10 here as well. Looking at the following chart, you'll see that the "10+" rating indicates an all-out sprint—equal to my old descriptor of "as fast as you would go if you were saving your child from an oncoming car." To provide a rough guide, the figures below match exertion levels with the proper score on Borg's revised scale.

Exertion rating

	10+	As hard as you can go; sprint-away-from-danger pace
Almost max	10	
	9	Starting to gasp for breath; unable to speak
	8	
Very heavy	7	Rapid forceful breathing and not wanting to talk
	6	
Heavy	5	Starting to breathe hard and get uncomfortable
	4	
Moderate	3	Deeper breathing but could speak in full sentences
Light	2	
Very light	1	Could easily hold this pace and talk for hours
Nothing at all	0	

What's with the Ascending Exertion Ratings for Some Workouts?

Because I perform most of my workouts on my home exercise bike—a Life Fitness 95Ci, to be specific—I tend to base my own sprint workouts on *absolute* workloads, which tie the outputs to a specific workload that I have to maintain. My bike allows me to plug in a certain power setting—say, 250 watts. And then I'll do five intervals of five minutes apiece with a minute of recovery. My heart rate climbs throughout the five sprints—and I know that last interval is going to be a beast. I like it because the challenge increases over the course of the workout. When I finish, I know I've given my all. I'm really spent—and I feel great.

Which brings us to one facet of the workouts you may already have noticed if you skipped ahead. Some of the exertion ratings increase through the course of the workouts. That's because I'm basing the workouts on protocols featured in the actual scientific studies that established the remarkable potency of high-intensity interval training. Some workouts are based on sprints with near-maximum or maximum efforts. In *those* cases, the person's power output or actual speed declines with each successive sprint—because the all-out effort of the previous sprint is tiring and makes less energy available for the next sprint.

However, some of these workouts are based on studies that required the subjects to generate the *same* power output with each sprint, regardless of whether it was the first or the last rep. *These* workouts tend to feature sprints that are intense, but something less than all-out. So, for example, in the study that established the potency of the Ten by One protocol, we asked the subjects to pedal at the same power output all the way through the workout. And to do that, they had to work a little harder each time. (The analogy holds for other types of interval training. For example, if you're a runner, envision being asked to run up the same hill at the same speed ten times in a row—of course, you would have to exert more effort to achieve the same speed through each successive hill run.)

Hence, the increasing exertion rating for the workouts that are intense but not all-out. Because we're citing studies that scientists conducted with absolute workloads, the exertion level would have increased throughout them. And to be sure you're getting the same benefits as the study subjects, you should increase your exertion throughout the workout as well.

A Few Quick Words About Calories

Many exercise books note the calories burned in a given workout. So do exercise bikes. The one in my basement even takes into account variables like the workload setting and duration, as well as my weight and age. But the "calories burned" measure is really just an estimate—an average for people *like* me. It's not an exact measure because no exercise bike can know every one of the variables that influence the number of calories burned. So many factors influence that statistic, including your own genetic makeup.

Also, we intend these exercises to be templates applicable to just about any aerobic exercise, from cycling and running to repeating burpees and stair climbing. The number of calories burned also changes depending on the type of exercise you're doing. And finally, there's the notion of *afterburn*—a word that trainers often use to describe the increased number of calories your body expends while recovering from the exercise. The more technical term for the phenomenon is "excess post-exercise oxygen consumption," or "EPOC," and it is influenced by the intensity of the exercise you've just done. The more intense the exercise, the higher the number of calories consumed in the afterburn. (We'll discuss EPOC and intense exercise more in chapter eight.)

The point is, we can't provide you with the exact number of calories burned during these workouts because it depends on the actual workload as well as your own physiology. Suffice it to say that when you factor in the afterburn, all the workouts described in this chapter consume more calories than an equivalent period of traditional steady-state exercise. For some of the workouts, we have verified this directly in the lab. For example, the Ten by One elicits an increase in energy expenditure (that is, calories burned over twenty-four hours) that is similar to a bout of moderate-intensity continuous exercise lasting about twice as long.

Modifying Your Interval Workout

The following workouts are only suggestions. Feel free to change them any way you see fit. That's the beauty of interval workouts. They can be variable to a nearly infinite degree. One thing I'd suggest considering is figuring out a way to get comfortable incorporating some resistance training into your workouts. That's because resistance training becomes increasingly important the older you get. It's the sort of exercise that builds strength; it encompasses weightlifting, bodyweight movements like push-ups and pull-ups, kettlebell training, and workouts using the Universal machines you see in fitness centers.

It's well established that combining aerobic and resistance training reduces body fat better than aerobic exercise alone.

The resistance training also combats the muscle-wasting effects of aging. And it makes you look better.

Happily, many interval-training protocols are well suited to incorporate resistance training. Several different approaches exist—and each one tends to be hyperefficient, because it works a lot of different physiological systems in a short amount of time. One involves performing bodyweight exercises during the "rest" portions. So in the four and a half minutes of rest specified between thirty-second sprints in the Wingate Classic, for example, you might perform squats, burpees, push-ups, or pull-ups. The trouble with this approach is that the resistance training tires you out, so you have less energy available to devote to performing your sprints.

The way that I prefer, which also happens to be the way that many personal trainers do things, involves bodyweight exercises, or resistance-training movements in which the body is the "weight" lifted. Whether we're talking push-ups, pull-ups, burpees, squats, or the dozens of other variations, bodyweight exercises work many different muscle groups, which means they also elevate the heart rate. Performing them quickly exercises the aerobic system—which in turn means that bodyweight exercise can be performed as interval-training sprints.

So to incorporate resistance training into a sprint workout, simply swap out the run, swim, or cycling activity with any bodyweight exercise that elevates the heart rate to a similar extent. The trick is to avoid workouts with long-duration sprints, because few people can perform bodyweight exercises for any significant length of time. I wouldn't try to adapt the Norwegian workout to a bodyweight approach, for example, because the four-minute-long intervals will be too tough if you do push-ups for your sprint intervals.

Among the protocols in chapters six and seven, the Fat Burner and the Tabata Classic both work well with bodyweight sprints, as does the Ten by One. But feel free to experiment with the protocols and customize them to your particular needs. I have a favorite bodyweight-sprint workout that involves ten minutes' worth of thirty-second-on, thirty-second-off intervals. The first thirty-second work interval is a warm-up exercise of jumping jacks. The second "on" interval is push-ups. The third is pull-ups from a bar, and the fourth is squats. Then repeat the cycles of push-ups, pull-ups, and squats until you've conducted three sets of each, for a total workout duration of ten minutes. It's fun, and super time-efficient because I'm getting in my daily resistance and aerobic training in less time than it takes to walk the dog.

Team sports are another way to incorporate interval workouts into your fitness routine. My family is crazy about team sports. I play hockey once a week. My wife plays both hockey and soccer, usually on multiple teams a season, and both my sons play competitive hockey. Such experiences have provided me with many opportunities to observe how those playing team sports can incorporate interval-based physical activity without actually performing a specific interval workout.

Think about it: In hockey, on a team with three full lines, a typical shift is thirty to forty-five seconds long. The shift amounts to a sprint—a nearly all-out burst of activity that leaves me huffing for breath, and my heart pounding, fifteen to twenty times a game. Hockey is a great interval workout.

So is soccer, with its breakaways and runs into space; players can jog more than six miles in a professional game. Full-court basketball qualifies as an interval workout, too. As do Ultimate Frisbee and touch football. In fact, many team sports amount to the best kind of sprint sessions. Conducted with friends, at near all-out intensities, they likely provide their participants with plenty of benefits.

But how much? A fellow academic from my hockey group, McMaster's Peter Kitchen, coauthored a 2016 study that found that male recreational hockey players ages thirty-five and over tend to have lower rates of diabetes, high blood pressure, and heart disease than do other physically active males.

Similarly, a 2010 paper revealed that recreational soccer players took part in intense intervals, with their heart rates averaging higher than 90 percent of their maximum for 20 percent of the game, leading to increases in muscle mass and cardiorespiratory fitness that were greater than those in a comparative group of endurance runners.

The point? If you're looking for a fun, social alternative to interval workouts that nevertheless provides many of the same benefits, consider playing team sports.

How to Approach the Workouts

All the workouts described in this chapter are based on formats used in actual scientific studies. Some of these studies featured warm-ups and cool-downs that ranged from five to ten minutes long, whereas other studies did not explicitly describe such details. Most people will do fine with a three-minute warm-up and a two-minute cool-down, so in the interests of time-efficiency and consistency, that's generally what is specified here. As you will see from the descriptions, certain workouts are more suited for certain types of people—whether beginners or those who have progressed further in their training. If you opt to set yourself on a path toward vigorous exercise, here's how I'd approach things:

Step 1: First, always check with a physician before starting or changing an exercise routine. Once you've received the all clear from your doctor . . .

Step 2: If you're out of shape, don't try to be a hero. Mitigate the low risk that exists by starting with easy workouts and then working your way up to tougher ones. Don't begin with all-out sprints. Instead, try an interval-walking program and move gradually to more intense workouts like the Ten by One and the 10-20-30 (both of which you can find in this chapter). We call these intense but submaximal protocols, because while the formats ask the participants to work out hard, the protocols don't request all-out intensities.

Step 3: Only when you're comfortable with variations of submaximal interval protocols should you consider moving up to the expert workouts—the bodyweight circuit-training formats like the Tabata Bodyweight or Go-To workouts in chapter seven. Same goes for the really potent, ultratime-efficient all-out workouts such as the One-Minute Workout and the Wingate Classic. I'll say it again: These are not for beginners—work up to them, and once you're ready, feel proud that you're in-shape enough to enjoy the benefits of the most potent, time-efficient exercises available.

The Workouts

The Beginner

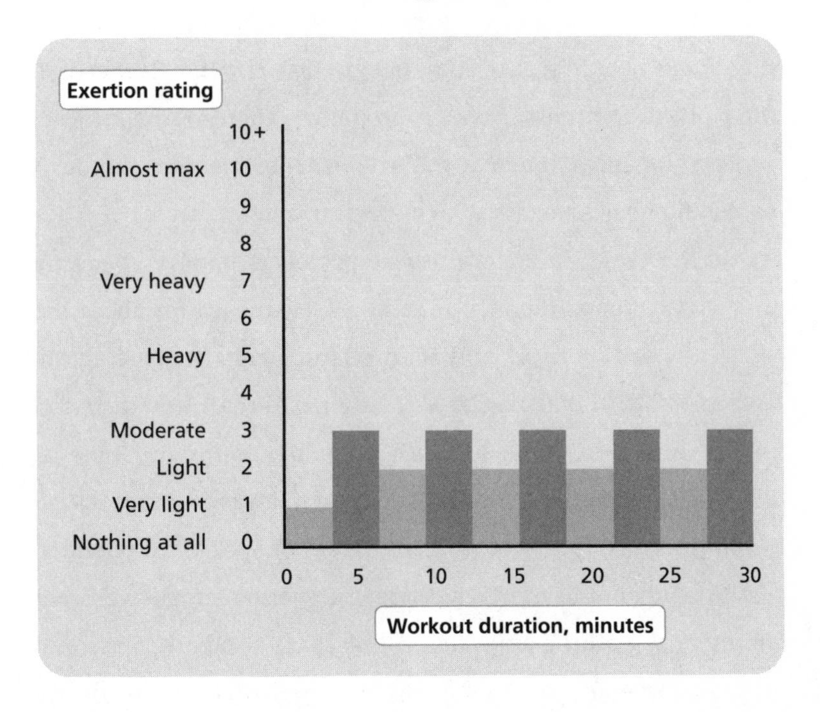

HIIT sounds daunting—like it's available only to those who are ultrafit. But even if your only physical activity is walking, you, too, can benefit from an interval approach.

Peak Intensity • 3

Duration • 30 minutes

The Evidence • Walking is the best medicine, according to many doctors. It's easy, convenient, and cheap. The problem is that some people's pace may not be fast enough to increase their physical fitness. So a group out of Japan's Shinshu University pioneered interval walking for older people who don't do much physical activity, kick-starting a whole series of studies on the topic. In general, these protocols involve speeding up the pace for around three minutes, easing off for about the same amount of time, and then pushing up the pace again. Compared with walking at a steady pace, the interval-based routine has been shown to result in much larger improvements in cardiorespiratory fitness and much larger decreases in blood pressure for those who are out of shape. More recently, a 2013 study out of the University of Copenhagen showed that an interval-walking approach cut fat mass and body mass and improved the ability of people with type 2 diabetes to control their blood sugar, while continuous walking did not. The interval-walking protocol we've used here is drawn from a 2010 joint effort out of the Mayo Clinic and Shinshu University. That study showed that three months of interval walking repeated four times a week increased cardiorespiratory fitness by more than 25 percent and spurred a 6 percent reduction in systolic blood pressure—the important first, higher number of any blood pressure reading. The Mayo Clinic–Shinshu University study subjects averaged 34-minute-long bouts of interval walking. I've capped the workout at 30 minutes here, but feel free to go longer for incrementally additional benefits.

Who Should Do It? • The Mayo-Shinshu study featured out-of-shape men and women in their mid-fifties. Previous studies have used this format on people in their late seventies who were so sedentary that three minutes of fast walking was the longest they could manage in a single effort. So this program could be well suited to seniors or really anyone just beginning an exercise regimen.

THE WORKOUT

1. Warm up by walking at an intensity of 1 for 3 minutes.
2. Increase your effort to intensity 3, so you are breathing deeply but could still maintain a conversation. Hold that pace for 3 minutes.
3. Ease back to intensity 2 for 3 minutes.
4. Repeat steps 1 to 3 for a total duration of 30 minutes.
5. Don't feel bad if you can't manage 30 minutes of interval walking right away—start with as many repeats as you can manage and then work your way up.

2

Basic Training

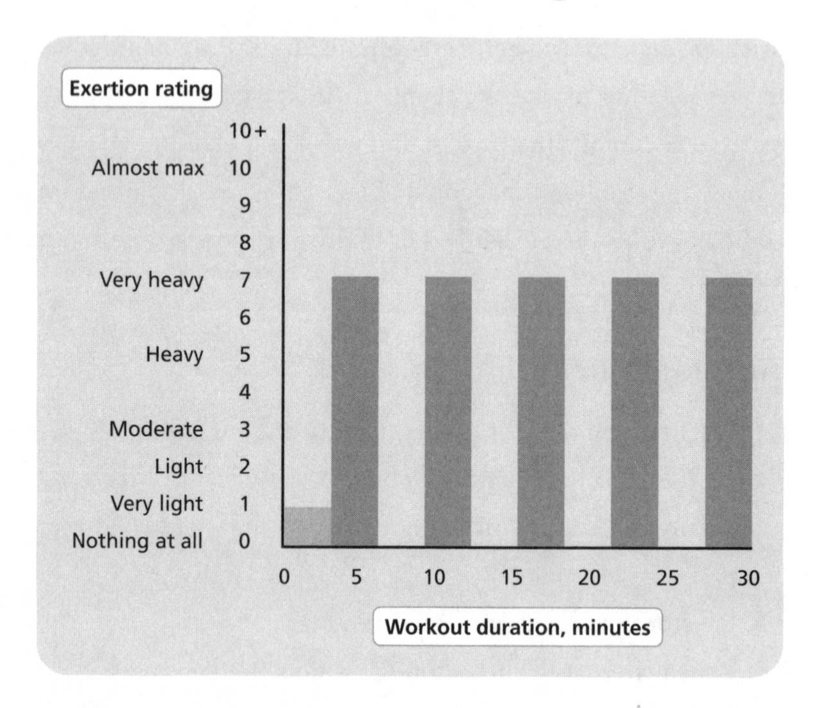

Based on a protocol from my late mentor, the legendary Scandinavian physiologist Bengt Saltin, this workout was featured in a 1973 study and is a progression from interval walking that remains based on 3-minute intervals. The long and comparatively hard sprints may be difficult for beginners but are a potent way to quickly boost cardiorespiratory fitness, enabling longer and faster aerobic activity. The original study protocol did not include a specific cool-down, and in the interest of time efficiency, I have not

included one here. But it never hurts to cool down at an easy pace for a couple of minutes.

Peak Intensity • 7

Duration • 30 minutes

The Evidence • In the early 1970s, Bengt Saltin and his coauthors recruited three platoons of young male conscripts starting basic training in the First Swedish Communications Regiment. Rather than putting them on the sort of calisthenic and running program common to military boot camps around the world, Saltin had one platoon run 15 minutes a day in 15-second-on, 15-second-off bursts. The two other platoons ran in cycles of 3 minutes on, 3 minutes off—but one of these platoons trained three times a week and the other trained five times a week. The biggest boost in cardiorespiratory fitness, with a 20 percent increase in just a month, was in the platoon that trained for 3 minutes on, 3 minutes off, three times a week—a basic and easy-to-remember protocol that can easily be done at a track or on an exercise bike.

Who Should Do It? • Saltin and his coauthors used healthy but untrained male conscripts in the Swedish military, with an average age of about twenty-one. While I wouldn't suggest 3-minute-long intervals for anyone just starting out with exercising, the protocol should benefit healthy people in the early to middle stages of training who wish to break through a performance plateau or quickly boost their cardiorespiratory fitness.

THE WORKOUT

1. Warm up at a very light intensity of 1 for 3 minutes.
2. Go hard up to an intensity of 7—where breathing is rapid and forceful and you're unable to talk—and keep it up for 3 minutes.
3. Rest for 3 minutes.
4. Repeat with intervals of 3 minutes on, 3 minutes off until you've completed a total of 5 sprint intervals.

The Norwegian

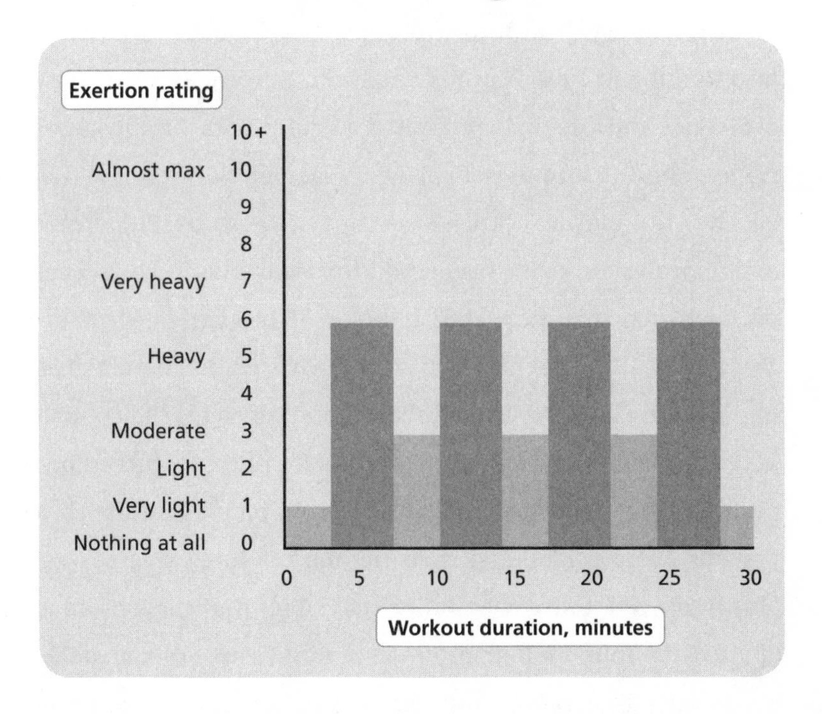

I like this one because dozens of studies have applied it to populations that others might consider too risky for HIIT, such as those with cardiovascular disease, type 2 diabetes, or metabolic syndrome. It's not just young, healthy people who are doing the hard intervals. Note that the original workouts were based on running or involved incline walking on a treadmill, but intervals could include brisk uphill walking or cycling, or many other modes of exercise.

Peak Intensity • 6

Duration • 30 minutes

The Evidence • This protocol is based on one of the most established interval-training protocols. The format first came to my attention in 2001. A group out of the Norwegian University of Science and Technology applied it to elite male junior soccer players. Just 16 minutes of sprints twice per week for 8 weeks boosted the players' cardiorespiratory fitness by 11 percent and, more remarkably, triggered improvements in key soccer performance indicators: the number of in-game sprints increased by 100 percent, the distance covered during matches improved by 20 percent, and the number of times the trained players touched the ball improved by 24 percent. Incredibly, performance indicators that didn't have much to do with a player's overall fitness, such as number of successful passes, also improved. Since then other academics and coaches have applied the four-by-four protocol to numerous other populations, including those with cardiovascular disease and the overweight. For example, a 2008 study conducted out of Ulrik Wisløff's lab at the same Norwegian institution applied the protocol to a group of sedentary middle-age men and women with metabolic syndrome, the precursor to heart disease, stroke, and diabetes. Repeated three times a week for 16 weeks, the format was able to actually reverse the symptoms of metabolic syndrome. In fact, the interval group increased their VO_{2max}, that critical marker of health, by 35 percent, more than twice as much as a group that performed continuous moderate exercise.

Who Should Do It? • This protocol has been applied to everyone from trained athletes to cardiac rehabilitation patients, although, to me, the most intriguing benefits happen in sedentary people who want to reverse the risk of metabolic syndrome.

THE WORKOUT

1. Conduct a 3-minute warm-up at an exertion level of intensity 1.
2. Go hard for 4 minutes at intensity 6.
3. Go easy at intensity 3 for 3 minutes.
4. Sprint for 4 minutes at intensity 6, and then go easy at intensity 3 for 3 minutes, repeating the cycle until you've completed 4 intervals of 4 minutes each at intensity 6.
5. After the 4th interval, recover for 2 minutes at intensity 1, for a total workout duration of 30 minutes.

The 10-20-30

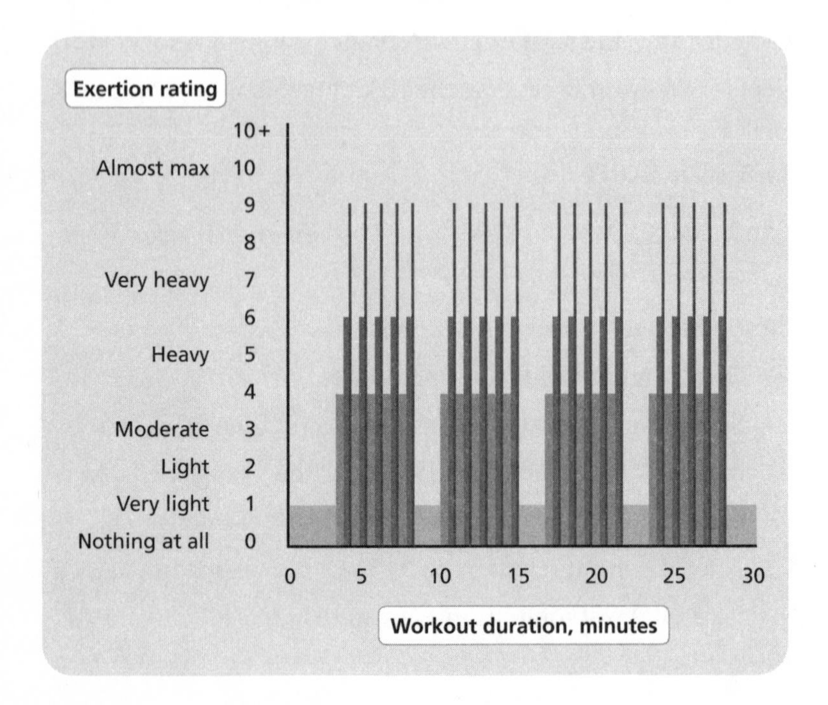

In 2012, some of my former colleagues at the University of Copenhagen published a study that examined how incorporating interval training would affect fit, trained runners. The protocol they used looks complex but ends up being pretty simple once you get the hang of it. It's also time-efficient, and so much fun that Gretchen Reynolds at the New York Times' Well Blog calls it her favorite workout. The protocol has the added benefit of triggering some remarkable benefits even in fit study subjects.

Peak Intensity • 9

Duration • 30 minutes

The Evidence • Going all out boosts cardiovascular performance—we know that. It also hurts. So physiologists at the University of Copenhagen wondered: Could fit runners experience performance benefits if they *didn't* go all out? To discover the answer, they took a group of 16 male and female runners who ran regularly—two to four times a week, on average covering about 17 miles in 2 hours and 15 minutes. Eight of them kept running as usual. For the other eight, the Danish scientists created an unusual protocol that nevertheless cut the interval group's training volume by 54 percent. After 7 weeks spent conducting the interval protocol three times a week, the scientists compared the two groups of runners. The endurance runners who had continued their training as normal hadn't changed anything. Those in the sprint group, who had conducted the 10-20-30 workout, had increased their VO_{2max} by 4 percent. They improved their times in the 1,500-meter run by 21 seconds (6 percent) and in the 5-kilometer run by 48 seconds (4 percent); decreased their systolic blood pressure by 5 mmHg; and also cut their counts for the bad type of cholesterol, known as low-density lipoprotein cholesterol. Not bad for a group of already-trained athletes doing about half the exercise that they had before.

Who Should Do It? • The Danish scientists conducted their study on trained runners with a mean age of approximately

34 who, before the study began, could run 5 kilometers in a mean time of about 23 minutes—quite a decent time. But the protocol seems likely to boost cardiorespiratory fitness in anyone who tries it.

THE WORKOUT

1. Warm up for 3 minutes.
2. Go for 30 seconds at intensity 4.
3. Go for 20 seconds at intensity 6.
4. Go for 10 seconds at intensity 9.
5. Repeat the cycle four additional times, so that you run intervals for a total of 5 minutes.
6. Rest with a light activity for 2 minutes.
7. If you're able, repeat with 3 more 10-20-30 sprint blocks.
8. Cool down for 2 minutes, so that the whole workout has taken about 30 minutes.

The Fat Burner

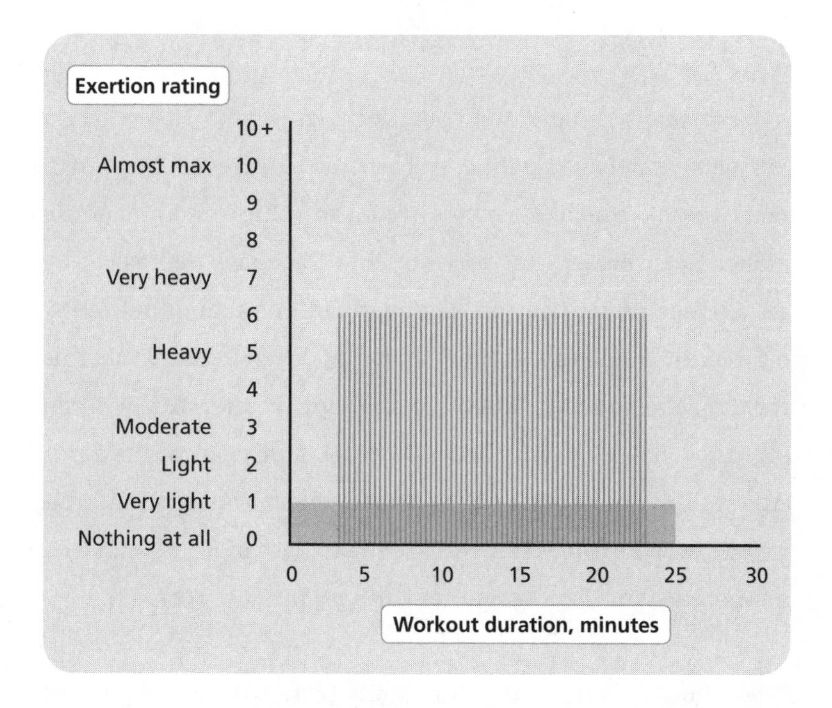

Noting the obesity epidemic, and the fact that many exercise programs lead to little or no fat loss, Australian scientists set out to design a protocol expressly for inactive people that targets fat loss. This is the result: a series of micro-intervals that are superior to conventional steady-state exercise at reducing fat. Another bonus? Recent research suggests people prefer to perform lots of shorter intervals rather than fewer longer intervals, and this study fits the bill.

Peak Intensity • 6

Duration • 25 minutes

The Evidence • The 2012 Australian study on which this protocol is based had subjects conduct the workout three times a week for 12 weeks, resulting in a 15 percent increase in cardiorespiratory fitness and some intriguing body-composition changes. Participants had a 3.3-pound decrease in overall bodyweight, which is really difficult to achieve with exercise alone. Even better, the subjects' total fat mass decreased by an average of 4.4 pounds. The workout cut abdominal fat by 6.6 percent and trunk fat by 8.4 percent. Keep in mind that this format relies on comparatively easy sprint intervals, with an intensity of only 6. An earlier 15-week study out of the same Australian lab increased the sprint intensity to an all-out rating of 10+, resulting in a decrease in fat mass of 5.5 pounds and an increase in cardiorespiratory fitness by 24 percent.

Who Should Do It? • The 2012 study featured overweight and inactive young men, while the earlier, more-intense study featured inactive young women whose body-fat percentage was more than 30 percent. But it's likely to benefit most healthy populations who feel up to trying it. Don't be scared off by how extreme this workout looks on the graph; with a comparatively easy sprint-intensity level of just 6, it really isn't that bad. Plus, the sprints are short, at 8 seconds, and the recovery periods of 12 seconds help a lot. Even if you push the intensity levels, this one is less painful and more accessible than you might think on first impression.

THE WORKOUT

1. The key to this protocol is figuring out a convenient way to keep track of the 8-seconds-on, 12-seconds-off format. I suggest staging the workout on an exercise bike, for cyclists, or at a track, for runners—and using a traditional stopwatch or an interval-training timer app that can be programmed to provide stop and start alerts. Or if you're not technologically minded, you could always just count off the seconds in your head.
2. Once you've got all that sorted, perform an easy warm-up for 3 minutes.
3. Go hard at intensity 6 for 8 seconds.
4. Rest for 12 seconds.
5. Repeat as many times as possible to a maximum of 60 times, or 20 minutes of intervals.
6. Perform an easy cool-down of 2 minutes.

6

The Wingate Classic

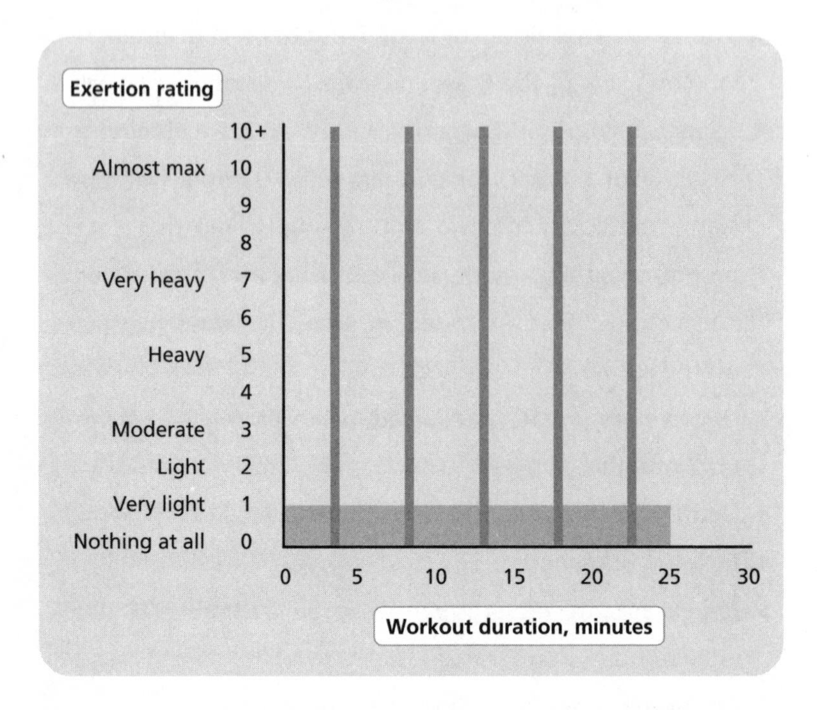

This is the original protocol that generated so much media attention in 2005 and again in 2008 when my lab showed that the workout could provide the same benefits as ten times as much conventional steady-state endurance exercise. So long as you're able to go all out, the protocol remains a potent and efficient tool to become fit—and stay there. Don't worry too much about what you do between the sprints, although you shouldn't feel like doing much if you are pushing hard enough. The key to this workout is

the intensity of the interval, which should be as hard as possible.

Peak Intensity • 10+

Duration • 25 minutes

The Evidence • When it comes to boosting fitness, there's something remarkably potent about going all out—and this is the protocol that helped us grasp that. We based it on repeats of the Wingate test, a 30-second all-out sprint on a stationary bike. It's exhausting—and remarkably powerful. The training protocol features a series of five 30-second all-out sprints, a total of just 2.5 minutes of hard exercise per day. In our study, we had our subjects repeat the protocol three times a week, amounting to a weekly time commitment of just 1.5 hours, and less than 10 minutes of hard exercise a week. After 6 weeks, we compared the sprint group's benefits with those experienced by a group that exercised continuously at a moderate intensity five times a week for a total of 4.5 hours a week, also for 6 weeks. The sprint subjects either equaled or exceeded the conventional exercisers in their improvements in aerobic capacity, muscle endurance, and the ability to burn fat. A remarkable result, considering the sprint group spent a third of the time exercising.

Who Should Do It? • Wingates aren't for everyone—nor is this protocol. We originally conducted the study on fit young men in their mid-twenties. Since then, other studies have shown the protocol improves a number of metabolic and vascular risk factors in overweight and obese adults. (If you're new to aerobic workouts, I'd suggest trying the Wingate Classic only after mastering easier interval workouts, such as the Beginner and the Ten by One.)

THE WORKOUT

1. Warm up for 3 minutes.
2. Blast through a 30-second sprint with an all-out maximal effort at an intensity level of 10+.
3. Recover for 4.5 minutes with some very light exercise.
4. Blast through another all-out 30-second sprint followed by 4.5 minutes of light exercise until you've completed 5 sprints.
5. Cool down with light exercise for 2 minutes for a total duration of 25 minutes and 30 seconds.

The Ten by One

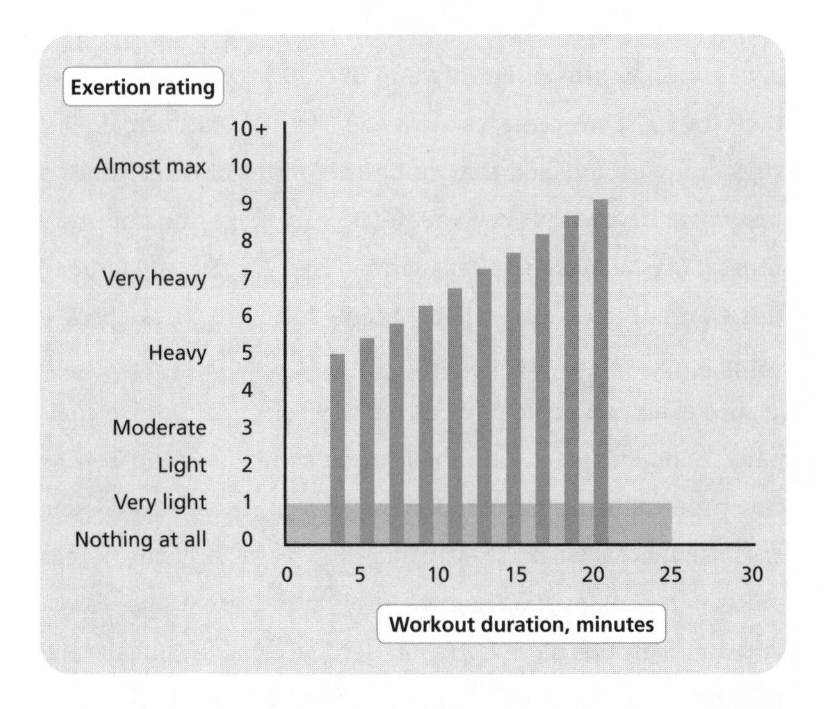

All-out Wingates aren't for everyone. So my lab came up with a less-intense version that nevertheless provides similar fitness benefits to those of the Wingate Classic workout. It's still time-efficient but more suitable for sedentary individuals beginning an exercise regimen, including those who are overweight. The minute-on, minute-off format is also easy to remember.

Peak Intensity • 9

Duration • 25 minutes

The Evidence • My lab began working around 2010 on an easier version of the Wingate Classic. Our first modified protocols toyed with lengths of sprints and overall power outputs, and then, in 2013, we came up with the Ten by One format. Our 2013 study established that doing this protocol three times a week over 6 weeks reduced body-fat percentage and improved fitness. My McMaster colleague Maureen MacDonald applied this workout to cardiac rehab patients in their sixties—people using exercise to treat cardiovascular disease. With both groups exercising twice a week for 12 weeks, MacDonald compared a high-intensity interval group conducting the Ten by One protocol to a group conducting 50 minutes of steady-state exercise twice a week. The sprint group was exercising half as much per week as the steady-state group. Despite the difference, both groups increased their aerobic capacity and arterial health to a similar extent—a promising result, considering that many cardiac rehab patients cite lack of time as a key reason for not adhering to their cardiologist's prescription to exercise.

Who Should Do It? • My lab's 2013 study tested this format on sedentary overweight women in their mid to late twenties. MacDonald's study applied the protocol to people with cardiovascular disease in their sixties. We have also shown it is effective to rapidly reduce blood sugar levels in older individuals with type 2 diabetes. But the format is fun and easy and could be used by virtually anyone healthy enough to conduct intense exercise.

THE WORKOUT

1. Warm up with light activity for 3 minutes.
2. Conduct your first sprint at intensity 5 for 1 minute.
3. Rest for 1 minute of recovery.
4. Conduct your second sprint for 1 minute at a slightly higher intensity level than the previous one.
5. Continue the cycle until you've done 10 sprints, with the final sprint conducted at intensity 9.
6. Cool down with light activity for 2 minutes, for a total duration of 24 minutes.

The Midwestern

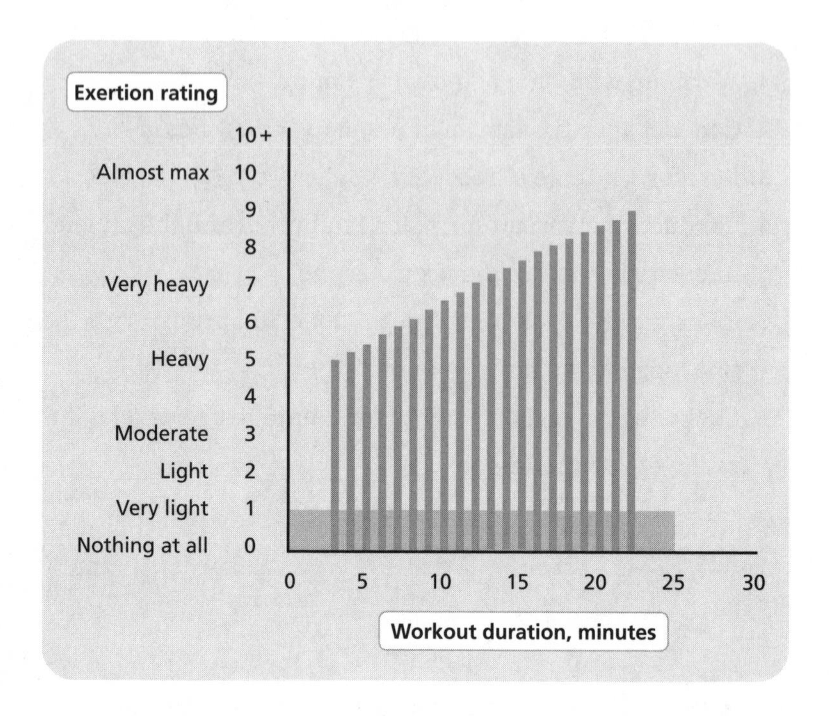

Sure, the Wingate Classic uses 30-second sprints, but that doesn't mean that every sprint needs to be all-out or painful. Reduce the intensity to a more bearable amount and increase the number and frequency of sprints and you have the Midwestern, an easy-to-remember, easy-to-customize, and fun protocol with some potent properties of its own.

Peak Intensity • 9

Duration • 25 minutes

The Evidence • Early in the 1990s, when few people were interested in interval training, a group of researchers at the University of Illinois at Chicago took a dozen untrained men and women in their late twenties and put them through an early comparative study that set intervals against continuous exercise. After 8 weeks, the continuous training group hadn't significantly improved their cardiorespiratory fitness, while the interval-training group's increased by 16 percent. The interesting thing about the resultant study? The researchers increased the resistance on the exercise bikes used for training so that the exercise intensity stayed the same while the subject's fitness improved. The interval-training group's fitness climbs high at a steep rate, while the continuous group's merely ambles upward.

Who Should Do It? • As with the Ten by One, this is an interval protocol that's fitting for those who are relatively early in their training programs and seeking a way to boost their fitness quickly and relatively painlessly, as submaximal 30-second sprints turn out to be quite bearable.

THE WORKOUT

1. Warm up at an intensity of 1 for 3 minutes.
2. Hike up the intensity to 5 for 30 seconds. By the end of the sprint, you should be breathing heavily.

3. Ease back down to an intensity of 1 for 30 seconds.
4. Repeat the pattern for as long as you can, gradually increasing your effort through each sprint, until you've conducted a total of 20 sprints. Those who can't complete all 20 shouldn't feel bad—if you stick with this protocol, you'll find that you'll make it all the way sooner than you expect.
5. Cool down at an intensity of 1 for 2 minutes.

Top Five Things to Remember When Designing Your Own HIIT Workouts

Let's face it—unless you have a coach or a trainer to keep you disciplined (or a squad of graduate students monitoring you for a scientific study), it's difficult to adhere to any single training protocol. Nor should you. Boredom sets in after a while, training plateaus, and everyone benefits from some variety now and again. Feel free to take elements of each of the above workouts to create your own exercise protocols. And when you do, keep this advice in mind:

1. There's No Free Lunch

It's a reality that you can't escape: If you want time-efficient workouts, you have to push yourself. To paraphrase something Izumi Tabata noted in his 1996 paper, the harder you go, the more time-efficient and potent the workout's fitness benefits. That doesn't mean you have to get out on the track or in the bike saddle and blast it every time you perform a workout. Just know that the harder you go, the greater the payoff in terms of health benefits and performance improvements.

2. Athletes Might Want to Opt for Longer Sprints . . .

Many exercise scientists and coaches believe that repeated hard intervals lasting between three and five minutes are most effective at boosting cardiorespiratory fitness. So if you want to be able to run, ride, or swim longer and faster without losing your breath, include some sessions that use this approach.

3. . . . But Short Sprints Are More Fun

A recent study conducted jointly between Mary Jung and Jon Little of the University of British Columbia's Okanagan campus, and researchers from the University of South Florida, found that previously sedentary overweight people preferred interval training that involved more frequent, shorter-duration efforts (as opposed to fewer sprints of longer duration). This was true even though each session involved the same total amount of work. The key message is that short intervals may be best to help keep newbies motivated.

4. Try Scattering Your Exercise Throughout the Day

We've grown accustomed to staging a single workout at a certain point in the day. But a 2014 study out of New Zealand showed that people with type 2 diabetes were better able to control their blood sugar if they distributed their sprints through the course of a day, rather than getting them done in a single block of time. This seems logical. As muscles work, they tend to use up their fuel stores of glycogen, and once those stores are depleted, the muscle tissue mops up sugars from the blood—reducing health risks like insulin resistance. The takeaway? Periodic bouts of exercise conducted throughout the day may be better for you, depending on your goals, than a single workout of long duration.

5. Variety Is the Spice of Life

I can't stress this point enough: one of the best aspects of interval training is the infinite variety. There's nothing wrong with traditional cardio, if you have the time, but there are only so many ways to do prolonged continuous exercise at a moderate pace. In contrast, the potential combinations afforded by intervals are endless. It's a way to exercise that's simply a lot more fun. So don't get wedded to any one of these interval workouts. Depending on your health and fitness level, allow yourself to pick and choose which one you feel like doing on any particular day—and feel free to design your workouts yourself.

CHAPTER SEVEN

How Low Can You Go?

Four Potent Microworkouts

WHILE I WAS PUTTING THE FINAL TOUCHES ON THIS book, I received an email from a guy named Andy Magness, an adventure racer who once was based in North Dakota and now lives on New Zealand's South Island. Early in his ultra-running career, Andy became accustomed to placing well in ultramarathons of fifty miles and adventure races that saw him spending anywhere from two to ten days traversing some of Earth's most forbidding territories. Then in 2007, he became a father and faced increasing work commitments—a situation very similar to the one I described at the beginning of this book. Andy was unable to afford the ten to fifteen hours a week he'd previously devoted to his training. The thing was, he still wanted to challenge himself with adventure races. The answer turned out to be high-intensity interval training. Today Andy still runs a half-dozen big races a year, from Wisconsin's 64-mile Frozen Otter Ultra Trek to the team-based, 6-day, 250-mile Abu Dhabi Adventure Challenge.

And yet Magness trains only thirty minutes a week.

How low can you go? People ask me that question all the time. I'll be up at the lectern for a speaking engagement at a conference. I'll finish my talk and open the floor for questions, and often the first person to raise a hand will ask, "How little can you get away with?" Or some variation.

What they mean is, what's the minimum amount of exercise possible to do while still ensuring that you live a long and healthy life?

For ultramarathoners, Andy Magness likely is brushing up against the minimum possible. Google him—he's all over the web, and he has an e-book about his interval-based training methods. For the rest of us—the people who just want to get in shape and stay there—the answer, based on current science, is a minute. A minute of hard exercise. You sprint as hard as you can, for twenty seconds, and then repeat that twice more for a total of three sprints? Congratulations. You've just done the most potent workout available.

I feel confident saying a minute because we just published a study that showed people who did a minute of all-out exercise three times a week, within a total time commitment of 30 minutes a week, had the same improvement over three months as the people who did all the exercise specified by the public health guidelines. That is, 150 minutes a week of continuous, moderate exercise.

Here's why: Intensity is more important than duration. Relative to all sorts of health benefits, it is more time-efficient to exercise hard for a short amount of time than it is to exercise easy for a long amount of time.

The reason comes down to the metaphoric switch in the body that we explained in chapter three. The switch that activates the body's exercise benefits: more mitochondria in the muscles, greater pumping capacity in the heart, more malleable arteries. Recall, two ways to activate the switch exist. The *traditional* way is to exercise at a moderate intensity for a long time, at a steady pace. The progressive decrease in energy stores triggers all sorts of adaptations. The *longer* you exercise in this manner, the better the effect.

We know by now though that there's another, faster way to activate this switch. This way flips the switch by depleting the energy stores *quickly*. It's the *rate* at which you deplete your energy, rather than the energy store's absolute level. And the faster you deplete the energy stores, the better. You make the energy stores go down *really* fast? Then you get *a lot* of exercise effects. The most time-effective way to get exercise effects using *this* method is to exercise as hard as you're able. And it's best to repeat your burst a few times in a row—to do at least a couple of intervals.

To flip the switch this way, it matter less how *long* you exercise. What's more important is that when you *do* exercise, you go *hard*. Exercise hard enough, you deplete your energy stores fast enough, and you can get remarkable benefits in a fraction of the time.

People find this difficult to grasp at first, because it's so different from the way we've traditionally thought about exercise benefits. It's tricky—that a handful of hard sprints might benefit us as much as a long run or cycle. So I like to use an analogy. Think of the body's exercise-benefits trigger like a pilot on a transoceanic flight. Say it's a 747 flying from New York to London, and it's three hours out of JFK. Two fuel-related things could seriously concern the people flying the plane. One, the pilot gets concerned when the plane has no fuel left. Of course that's concerning, right? Because the plane isn't going to make it to Heathrow. The news inspires a flurry of activity on the plane. Passengers rush to their seats, air masks come down from the ceiling, that sort of thing.

Another fuel-related thing that will concern the pilot is a rapid *decrease* in the fuel supply. Like if the gauge goes from full to half full in just a minute or two. At that rate, the tank is going to be empty way before the plane ever touches down at Heathrow. That information inspires a flurry of activity on the plane, as well.

The human body's exercise regulator is like a pilot—except, when *it* gets concerned, it starts making changes. "This body has been working out for so long it's exhausting its fuel stores!" cries the exercise regulator. Or, "This body is working out so hard it's *rapidly depleting* its fuel stores!" Both things cause the exercise regulator to get concerned and initiate changes in the body to lessen the negative impact. The heart-strengthening, mitochondrial-boosting changes we've discussed. This is the process of *training*.

The following graph illustrates the spirit of this process. Essentially, the harder you exercise, the more benefit you're going to get per unit of time. It doesn't matter so much how *long* you exercise; rather, it matters how *hard* you push yourself when you do exercise. So if you don't have the hours for endurance training, you can still get the same benefits with *minutes* of short, hard bursts.

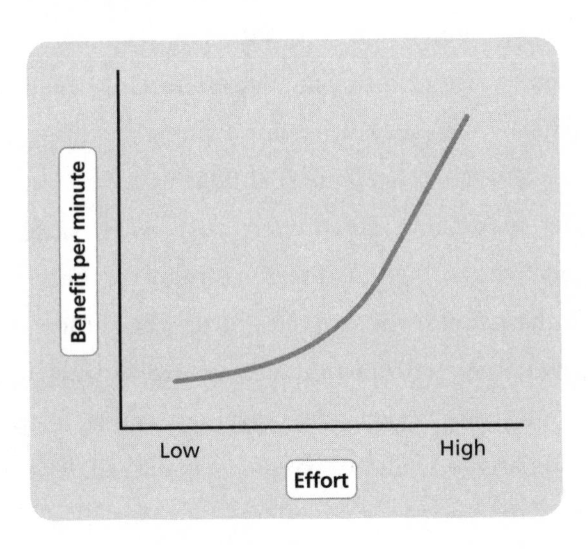

The Science Behind Intensity Over Duration

What started us on the intensity-over-duration path is something known as an epidemiological study—a complex term for something that's actually pretty simple in concept. In epidemiological studies, scientists follow a large number of people over time. Then they make comparisons between a person's lifestyle and health. And if you have enough people—say, thousands of them—then you can draw some conclusions: People who did *this* tended to live longer than people who did *that*.

The principle of *intensity over duration* has been established by numerous key studies released in recent years. Take the Copenhagen City Heart Study, which illustrated the principle in two different ways. It tracked nearly twenty thousand Copenhagen residents starting in 1976 and continuing for up to twenty-five years. By comparing the lifestyle information that the subjects provided with the causes of death over time, the investigators were able to establish the relative health benefits of certain lifestyle choices—for example, walking. As time went on, lead researcher Peter Schnohr and his team noticed something interesting about their subjects' walking habits. The people who walked a lot weren't living all that much longer than the people who walked a little bit. Rather, the clear finding was that those people who tended to walk fast lived longer. And it didn't matter whether they walked *a lot* or a little. The faster you walked, the longer you lived. The *distance* you walked didn't have much effect.

Five years later, further research revealed the same phenomenon for cycling. Since Copenhagen is one of the world's great cycling cities, a lot of people pedaled bikes as a key form of transportation. I can attest to this from the time I lived there working in Bengt Saltin's lab. Just as with the walking study, researchers found that it didn't matter how many miles the subjects cycled per day when it came to their health and longevity. Rather, what mattered was how *fast* they cycled when they did. If they regularly cycled hard, even for a short time, then the subjects tended to live longer. Men who cycled fast tended to live 5.3 years longer than those who cycled slowly; fast-cycling women tended to live 3.9 years longer than their slower-cycling counterparts. There was absolutely no relationship for the *duration* of cycling.

Other studies indicate a similar phenomenon. A 2011 Taiwanese study that appeared in the *Lancet* tracked more than four hundred thousand people, and it turned out that sixty minutes of moderate physical activity a day cut their overall risk of dying by about 25 percent—a remarkable figure. But those who did only fifteen minutes of physical activity a day also saw the same reduction—so long as they were exercising *vigorously*.

So how do you apply this principal of intensity over duration? That's what the rest of this chapter is about. I'm going to show you several ways that you can apply it to get greater benefits from exercise—as well as to save time, exercise more efficiently, and gain greater performance benefits when you do. What I'm hoping is that these principles will open up the benefits of exercise to proportionally more people.

Getting Down to One Minute

Mindful of the epidemiological evidence that showed intensity trumps duration, I sat down in 2014 with the graduate students who worked in my lab and devised our shortest, most potent workout to date. We were influenced by a 2011 UK study that addressed the concept of the minimal amount of exercise that could improve health. In that study, lead author Richard Metcalfe, then a graduate student at Scotland's Heriot-Watt University, took a look at my lab's Wingate Classic workout. That entails four to six repeats of thirty-second all-out sprints.

Going all-out for thirty seconds is pretty darn hard, Metcalfe thought. Recall *Esquire* writer A. J. Jacobs's description: "It feels like your legs are giving birth. It feels like you've got an eight-martini hangover in your calves." So Metcalfe wondered, is there a way to get *most* of the benefits of the Wingate Classic without so much pain?

Our own examination of that question led to the Ten by One protocol that has been so effective for so many different people. Factoring in warm-up and cool-down, our Ten by One takes twenty-five minutes. The UK researchers were looking for something even more time-efficient. They thought about the way that the benefits of intense exercise happen because hard physical activity recruits *all* the muscle's fibers, causing the fuel stores in the muscle cells to decrease quickly. Metcalfe and his colleagues realized that most of that decrease in fuel levels happens during the first fifteen seconds of the sprint. In fact, after this time, metabolic by-products start to accumulate that actually inhibit energy provision—that's one of the reasons sprinting gets painful after some time. So what about all-out sprints that were *twenty* seconds long? They'd be almost as effective as the thirty-second sprints, wouldn't they? Certainly, they would hurt less.

Metcalfe established a protocol that places only two sprints of twenty seconds in a workout that, including warm-up, recovery, and cool-down, lasts just ten minutes long. He called it reduced-exertion high-intensity training, or REHIT. In a selection of young, healthy but comparatively out-of-shape people, Metcalfe's twenty-second sprint protocols conducted three times a week over six weeks increased cardiorespiratory fitness by 15 percent in men and 12 percent in women. That's quite a result.

We were intrigued enough to conduct our own investigation into the benefits of twenty-second sprints. Another big influence was Tabata's 1996 protocol, which involves eight bursts for twenty seconds, except the rest period in Tabata is just ten seconds, for a total time of four minutes. People find Tabata workouts pretty killer by virtue of the minimal rest.

In our lab, we tried to come up with a protocol that involved *some* rest, but not *too* much, as it may not be necessary and detracts from the time efficiency, which was the whole point to start with. So when we designed our own study we used the same twenty-second sprints as Metcalfe and Tabata—with some modifications of our own.

First we considered the number of repeats. We definitely wanted to get *one* in. But come on. We're talking twenty seconds. Surely we could ask subjects to do more. Recall the two graphs I included in the book's first chapter—the way they demonstrated that the aerobic energy system is taxed more with each repeated sprint that you do. Then again, repeated sprints follow a law of diminishing returns. The more sprints you do, the less benefit you get from each additional one. So how many would maximize the benefit while minimizing the pain and the time?

We decided the answer was three repeats. We set our three repeats into a workout that lasted ten minutes, from start to finish. So a two-minute warm-up followed by a twenty-second sprint, repeated three times and followed by a three-minute cool-down. We had a group of subjects repeat the One-Minute Workout three times a week for six weeks. Our study also was more involved than Metcalfe's previous twenty-second studies, as we included additional measurements like muscle biopsies to examine changes in the subjects's muscles. The key findings were a 12 percent increase in cardiorespiratory fitness and a 7 percent decrease in resting arterial blood pressure. The study generated a flurry of media attention and was accessed more than fifty thousand times in the first year that it was published in the journal *PLoS ONE*.

Next, in an experiment that lasted twice as long, at twelve weeks, we compared a sprint group that did three minutes of hard exercise a week, set into a protocol that lasted thirty minutes a week, with a group that exercised for 150 minutes a week, as recommended by the guidelines.

Incredibly, the benefits turned out to be the same.

It was a stunning verification of the principle of *intensity over duration*. The improvement in cardiorespiratory fitness was the same in both groups—an increase of 19 percent. Body fat percentage decreased by 2 percent for both groups. Muscle mitochondrial content increased to a similar extent. And the improvement in how well the subjects could manage their blood sugar? The same.

That's right. It was possible for everyday, nonathletic sedentary individuals to derive the cardiorespiratory benefits of 150 minutes a week of traditional endurance training—three 50-minute sessions per week—with just a single minute's worth of hard exercise repeated three times per week. I've described the One-Minute Workout at the end of this chapter in more detail, in case you're ready to try it yourself.

Beyond the One-Minute Workout

The original One-Minute Workout study was performed on an exercise bicycle. Then we started thinking, well, how many people have convenient access to an exercise bicycle? Could we create a similar workout that's more practical for most people yet still as potent? This time my McMaster lab collaborated with a team out of Queen's University. We decided to investigate whether we could conduct a version of the One-Minute workout that could be performed on stairs.

After all, many of us work in office towers, and if we don't *work* in one, then maybe we live in a high-rise condo. Or spend time while traveling for business in hotels. Anyway, we took sedentary people from around the McMaster community and set them up in the university's six-story business building.

Some testing revealed that the average person could climb somewhere between four and five flights of stairs in twenty seconds. The participants climbed an average of around sixty steps during each bout. That became the basis of our workout; people simply walked around for the warm-up and then sprinted up stairs for twenty seconds. They walked back down the stairs, walked some more, then again climbed up the stairs as fast as they could. Next they followed one more repeat with a three-minute cool-down.

As I write this, the study hasn't yet been published. But the results suggest that it shouldn't matter whether one conducts the One-Minute Workout on stairs or an exercise bike.

We're also investigating one more flavor of this workout, to provide people with a more convenient option. Because, we realized, some people don't have exercise bikes or access to high-rise stairs. For them, we designed a version of the One-Minute protocol that uses just a conventional staircase, the sort that might be found in any two-story home across North America. Rather than a twenty-second sprint, it's based on a minute's worth of sprinting up and down the steps, repeated twice more within the space of ten minutes—I guess technically that means we should have called it the Three-Minute Workout. But the actual stair-climbing portion of the intervals lasts about twenty seconds. That is, each interval involves about twenty seconds of climbing and forty seconds of descending the stairs. The workout would have been particularly appropriate back in the days when people watched TV shows with commercials. You're watching a show, a commercial comes on, and you head for the stairs, sprinting upstairs and going downstairs and doing it again for all of a minute-long commercial break.

Using Ultralow-Volume Exercise

Different people will use the principle of intensity over duration in different ways. In general, three sorts of people are interested in cutting-edge exercise techniques—and each type can benefit from the science of ultralow-volume exercise. The first are the out-of-shape folks who want to get in shape—the group I call *unfit to fit*. The second are the athletes who are seeking a more time-efficient way to get performance benefits—or a potent way to break through a performance plateau. I call this group *fit to fitter*. The third group—which I belong to—are in-shape people with demanding careers who are fit enough but seeking time-efficient ways to use exercise to stave off the aging effects of inactivity. This group is called *maintaining fitness—or fighting aging*.

Unfit to Fit

You've already read about the remarkable benefits of interval training for sedentary people who want to get fit fast. But what's the least amount of exercise that will start people on the road to fitness? The point here is to get people exercising in a way that takes them through the gauntlet of pain experienced by anyone starting to exercise—in the least amount of discomfort. To get them fit, fast.

Bear in mind, too, that this isn't about going from fat to thin. That has more to do with what you put in your mouth—more on that in the next chapter. So this question is purely about how an out-of-shape person can start enjoying the benefits of exercise in the most time-efficient manner possible.

With the usual caveats, such as check with a doctor to make sure it's OK for you to exercise very intensely, one option is to start with a less-intense version of the One-Minute Workout. On stairs, on a bike, or by adapting the protocol to running in a field or on a trail—it doesn't really matter how you exert yourself. What matters is that you get out there and push a bit. Get out of your comfort zone. There are benefits to just a few minutes of exercise. Then, once you're feeling more capable, branch out to other workouts that you might find more interesting and challenging, like the Ten by One or, if you're looking at getting more buff, circuit-training protocols like the Tabata Bodyweight or the Go-To Workout, described in detail at the end of this chapter.

Fit to Fitter

Let's say you're already pretty fit—a recreational athlete, kind of a weekend-warrior type. But your enthusiasm is waning because you haven't experienced the sort of increases in fitness that happened when you first started working out.

Sprints can help.

Intense interval exercise is a time-efficient way to break through a performance plateau. The classic example of this is Roger Bannister studying at Oxford's medical school while training to break the four-minute mile—with all-out intervals.

More recently, a Danish physiologist named Jens Bangsbo looked at the problem of breaking through performance plateaus. As an assistant coach for Italy's Juventus Football Club, one of the world's best soccer teams, as well as the Danish national soccer team, he's had lots of experience working with professional athletes. Many of his studies entail replacing a portion of an athlete's regular training program with variations of intervals.

One of Bangsbo's studies saw him monitoring trained runners who typically did about fifty-five kilometers (around thirty-four miles) a week of roadwork. These athletes were already incorporating high-intensity intervals in their normal routine. Bangsbo had some of the subjects replace two of their weekly workouts with a series of all-out sprints. The sessions involved six to twelve sprints of thirty seconds each at near-maximal effort, with three minutes of recovery. The others maintained their normal training. After six to nine weeks of the intervention, the runners underwent muscle biopsies and a series of performance tests. The runners who'd persisted with their normal routine didn't improve at all; the sprinters improved their three-kilometer and ten-kilometer race times by around 3 percent—certainly enough to break through a performance plateau. And Bangsbo created even more benefits recently with a different protocol: the 10-20-30 workout that I described in the previous chapter.

If you're a highly trained endurance athlete, chances are you've already incorporated intervals in your training. But for the weekend warriors who haven't yet, intervals represent a potent way to blast through a plateau. Some critics of Bangsbo's study point out that some of his sprinters' benefits might have come from the tapering effects of lowered training volume—because trained runners who ease off on their workouts tend to perform better in races in the short term. Maybe. The fact is, trading regular endurance training for some form of sprint workout has been shown to substantially improve performance.

But let's say you're a recreational athlete who wants to boost your fitness *the most* in the shortest number of weeks. You have the time and the commitment. The answer is a forty-year-old program that's assumed near-legendary status among physiologists and coaches for the remarkable rate at which it's been able to boost cardiorespiratory fitness. The workout has created a template for the type of program intended to create the greatest VO_{2max} boost in the shortest number of weeks.

Named after the Washington University postdoctoral researcher who established it, Robert Hickson, the Hickson protocol features alternating days of interval and endurance training for six days a week over ten weeks. The interval days amount to six repeats of five-minute sprints on an exercise bike at VO_{2max}, which, for that duration of sprint, amounts to an intensity of about 9 on the exertion scale we have adopted in this book. Between the sprints, Hickson's subjects rested for two minutes with easy pedaling.

Endurance training days required running "as fast as possible" for thirty minutes the first week, thirty-five minutes the second week, and forty minutes for the remainder of the program.

Hickson noted in the study that most protocols that dumped so much training on previously sedentary individuals would render them sore and possibly injured. This one didn't injure its subjects, possibly because they alternated between the two very different activities of running and cycling. And after just ten weeks, Hickson's eight subjects had increased their cardiorespiratory fitness by an average of 44 percent—better than 4 percent a week.

It's not particularly time-efficient, and it's so tough that none of the subjects wanted to continue the training, but if you're focused on maximally increasing your cardiorespiratory fitness in a single season, the Hickson protocol is probably the best way to do it.

The final question is, what's the minimum level required to *maintain* that fitness boost? Drawing on the work of the Mayo Clinic's Michael Joyner, who coauthored a review on the topic of VO_{2max}-boosting workouts, it's apparently necessary to conduct two-thirds of the training volume of the Hickson protocol to maintain its gains. Happy training.

Maintaining Fitness—or Fighting Aging

What if much of the decline in function and strength that we associate with getting older doesn't actually have much to do with aging? What if, instead, the greater part of that age-associated deterioration actually has to do with *inactivity itself*? It's an idea that's gaining currency in exercise circles, reflected by a recent *Journal of Physiology* paper by a group out of King's College London. Lead-authored by Stephen Harridge, the paper points out that we associate inactivity with aging because so many older people fail to get much exercise. But what if we're mistaking cause for effect? What if it's not the *aging* that causes the inactivity, but the *inactivity* that causes the aging?

In that case, it would be possible to fight time's decline by maintaining high physical-activity levels. And a growing body of science suggests that it is possible to do that—to fight aging with lots of exercise.

That's what I'm trying to do, anyway. As I write this I'm approaching fifty, and I'm obsessed with exercise because I regard it as the most effective way to maintain health and well-being for many decades to come.

So what's the minimum level of activity required to do that?

The answer is, it depends.

Ulrik Wisløff led a study published a decade ago that makes the answer seem pretty easy. He and his team followed fifty-six thousand men and women for sixteen years. They found that a *single* weekly bout of high-intensity exercise was associated with a reduced risk of dying from cardiovascular disease.

We expect that, right? The stuff is potent. What was unexpected was that *more* exercise had no effect. Those who exercised once a week, for longer times, were about as likely to die from cardiovascular disease as those who exercised once a week for shorter durations. So were the people who exercised three or four more times—the point is, more frequently than once per week. To get the maximum effect of cardiovascular disease risk reduction, Wisløff's study found that all you have to do is exercise hard once per week.

The study results contradict the perception that exercising hard once per week is dangerous. Think of the classic middle-age guy who plays hockey or basketball once a week with his buddies. You occasionally see newspaper reports suggesting this is risky. To be sure, there are folks who suffer heart attacks after performing a bout of strenuous exercise. Probably, they hadn't been properly screened for risk. They might have been primed for a heart attack, which could have been brought on soon by a variety of activities or stressors. The scientific evidence suggests this once-a-week exertion via sport is likely beneficial in terms of an all-cause risk of mortality. Once again, the key message is to get checked by your doctor first.

I'm concerned about more than just my risk of dying from cardiovascular disease, though. I want to retain two things: my fitness and my strength—my ability to exert myself over time *and* my capability to complete tasks by lifting, pushing, or pulling. So I exercise for a half hour a day, at least, every day. I mainly alternate my days between body-weight-style resistance training and cycling, both performed in an interval manner for three days a week each, along with playing ice hockey with some buddies for an hour each week.

A typical time-efficient resistance-training workout sees me perform three exercises. The fastest way to trigger the strength-building benefits of resistance exercise is to go to failure on every set. Perform the exercise until you don't have the strength to do any more. It's challenging, sure, but that's what I do. I go down in my basement and do as many pull-ups as I can, until I can't do any more. Then I get down and do as many push-ups as I can until I can't do any more. Then I do it all over again with lunges. Each exercise takes me twenty to thirty seconds. I do have a power rack in my basement, a barbell, and some weights. This provides some options, and so I will typically vary things in a single workout—chin-ups, bench presses, and squats, for example. Once I've gone through them all, I start again, for a total of three sets each—every one to failure.

It doesn't take me very long, but it's pretty effective.

On my cardio days, I cycle on an exercise bike—just blast through some sprints. I'll select one of the workouts in this or the previous chapter. The Ten by One, say. A minute on, a minute off, for twenty minutes, and done. Or Bangsbo's 10-20-30. If I don't have that much time, I'll blast through the One-Minute Workout. Or if I'm *really* craving variety, I'll do the Fat Burner protocol of eight seconds on, twelve seconds off.

If I happen to be in a hotel that doesn't have an exercise bike in its workout area, I'll do almost anything to get my heart rate up. Remember, the principle of intensity over duration suggests that for most people, even a short burst at a high heart rate will help maintain or even increase your fitness. If it's a strength day, I'll do as much of the circuit routine as possible. Cardio? I'll go to check out the stairs, to see if its possible to do intervals there. Even with my bum left knee, which now makes running impossible, I find I can briskly climb a few flights of stairs without pain. Or I'll burst through a few sets of burpees—which brings us to the next section.

If You're Going to Do Only One Exercise . . .

Back in 2011, the *New York Times*' Gretchen Reynolds called me up and asked me the following question: If I were going to do only one exercise, which one would it be? Like if I had *time* to do only a single exercise. As in, which exercise is *the best* exercise? Similar to most academics who are forced into a corner, I attempted to reject the premise. "Trying to choose one is like trying to condense the entire field," I sputtered. Then I shrugged and provided Gretchen with my choice.

I said the best exercise was the burpee, the set of do-anywhere movements that some of us will remember from high school gym class. The first set of movements involves squatting down, placing your hands on the ground, and kicking both feet out behind you so you're in a plank. You can perform a push-up here if you like. Next, pull the feet back in so that you're crouching with your hands on the floor. Finally, leap up into the air and raise your hands above your head at the same time.

"Why the burpee?" Gretchen asked.

A couple of reasons. The biggest one is that the burpee simultaneously builds muscle and endurance, particularly when you incorporate a push-up into the movement. "But," I said, "it's hard to imagine most people enjoying it—or sticking with it for long."

In Manhattan, Joshua Spodek read the article and took it as a challenge. Spodek is an adjunct professor at New York University and a leadership consultant who teaches at Columbia. He's got a PhD in astrophysics as well as an MBA. He's the sort of guy who likes challenges. He's run marathons, swum across the Hudson River, and taken a cold shower every day for a month—just to see how it affected him. He competed at a national level in Ultimate Frisbee, and when he read Gretchen's article, he'd been looking for his next thing. He was up on his roof, having a couple of drinks with a friend, and soon after the two of them discussed the article, they both pledged to perform burpees every day for a month.

Spodek started the next day—December 22, 2011. The first time he did them, he made it up to ten before they exhausted him. The experience sold him on the exercise, which he saw as an efficient way to elevate the heart rate while also working in a resistance-training fix. The shoulders, chest, core, buttocks, quads, and calves—a properly done burpee works all of them. He also liked that the exercise didn't require any special equipment. "I thought, I'm going to do these forever," Spodek says. "They're everything I'm looking for."

At the end of the first month, Spodek had done three hundred burpees. After his repeats, he felt exhilarated. He developed a routine of doing them in the morning before his shower. His capacity improved to eleven, then twelve and thirteen, and then Spodek doubled his sets so that he was doing another bunch of burpees at night.

After a year, Spodek was up to two sets of twenty burpees a day. He was in his early forties, and he felt better than he had in years. So he kept at it—a set in the morning, a set in the evening. His shoulders filled out. His chest gained definition. "They put me in the best shape of my life," Spodek says.

The first year, he calculates, he'd done between nine thousand and ten thousand burpees. The next year he hit twenty thousand, then thirty thousand, and more recently, after five years of doing burpees and having increased his per-set repeats to twenty-six, with two sets a day, Spodek reckons he's squatted, pushed, and leapt his way up to seventy thousand burpees.

He's added to his twice-daily routine, too. Spodek found that the burpee repeats weren't working his back muscles because they didn't include a pull motion. One easy way to address this would be to install a pull-up bar somewhere—that's what I've done at my place. But Spodek's Manhattan apartment doesn't have any space for that. So after his burpees, Spodek sets himself up underneath his kitchen table so that he's hanging from his arms. He gets his heels up so that they're on a chair, so that he's hanging more or less horizontally underneath his kitchen table. Then he does a row—a reverse push-up. (Those looking to conduct an easier rowing motion can just leave their feet on the floor.) Spodek also stretches and conducts a series of ab workouts, including a fiendishly difficult maneuver called an L-sit that involves sitting on the floor with your legs straight out in front of you, using your hands to raise your bum off the floor, while trying to keep your heels off the floor, so that your body is at a right angle, almost in a pike position.

In a blog post, Spodek wrote about his burpee-based routine: "I've done them alone, with people, in public, indoors, outdoors, drunk, sober, hungry, full, early, late, happy, frustrated, and every way you can imagine feeling and being every day for a year. I've done them in New York City, Hollywood, North Korea, South Korea, China, Vietnam, Singapore, and the Philippines."

Recently, as I was finishing up this book, Spodek emailed me. "I've done them every day since soon after reading the article," he wrote. "I thought you might be interested in my experience, now in its fifth year, started from your story. I had never heard of burpees before . . . I figure if someone followed up a quote of mine like this, I'd want to hear back about it, too."

Thanks, Josh, for letting me know. I admire your dedication—and your discipline. What benefit is Spodek getting from his burpees? Well, he performs the vigorous movement quickly. The pace approaches all-out, and performing twenty-six amounts to a submaximal sprint of about ninety seconds, repeated twice a day. Factoring in the under-the-table rows, Spodek is building and maintaining strength in each of his major muscle groups. He's fighting aging and maintaining his cardiovascular fitness. Unlike Spodek, I could never do the same exercise routine every day, because I'd get bored—and besides, switching things up is good for the body. But if you're in a pinch, a burpee-based workout like Spodek's will provide much of the maintenance you need.

If You Really Don't Have Time, at Least Get One In

Finally, some advice for those who truly feel that they don't have time to exercise. In 2013 I coauthored a study out of Ulrik Wisløff's lab at the Norwegian University of Science and Technology. It tracked two groups of inactive but otherwise healthy overweight men who worked out three times a week for ten weeks. One group followed the Norwegian workout described in this chapter, performing four repeats of four-minute intervals each time they worked out; whereas the other group performed a single four-minute effort each time. The first group boosted their VO_{2max} by 13 percent, while the single-interval group improved it by 10 percent despite doing less than a quarter of the work. The point? If you're pressed for time, even getting in a single hard effort can go a long way toward boosting or maintaining your cardiovascular fitness.

Here, then, are a series of ultralow-volume workouts. These are the most potent exercise protocols science has yet developed. As with the protocols in the previous chapter, I've made some tweaks to the formats—standardizing warm-ups at three minutes and cool-downs at two minutes, for example. Feel free to customize them—and most important, have fun.

The Workouts

The One-Minute Workout

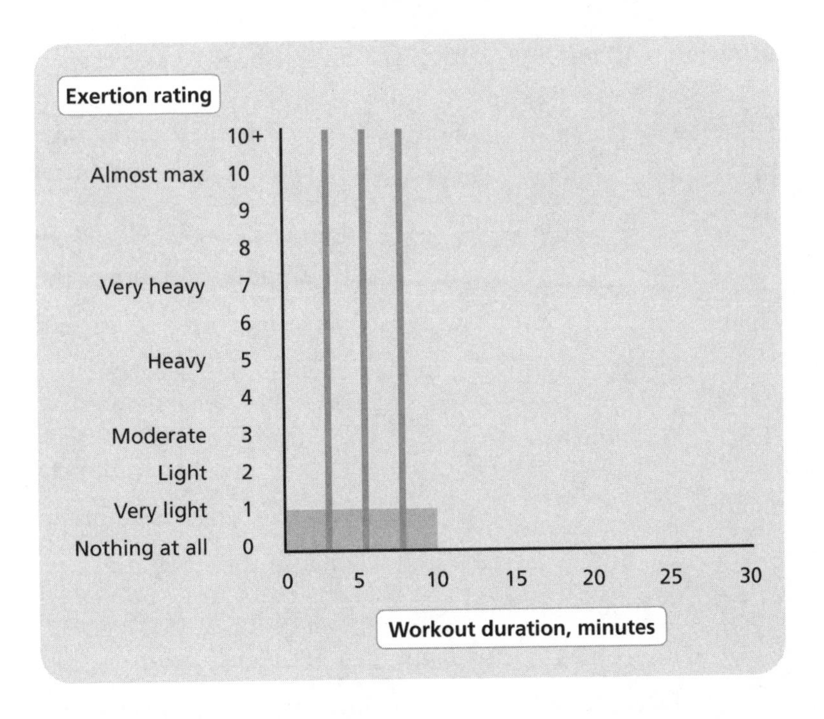

One of the most exciting areas of cutting-edge exercise physiology is the new science of ultralow-volume interval training. Influenced by a 2011 UK study investigating the minimal amount of exercise that could improve health, we developed our one-minute protocol for previously sedentary people who sought a time-efficient way to experience big improvements in overall fitness. How potent is this protocol? Our latest study shows that in just 10 minutes, with just a single minute's worth of hard exercise, it

provides the benefits of 50 minutes of traditional endurance training.

Peak Intensity • 10+

Duration • 10 minutes, with just 1 minute of hard exercise

The Evidence • Each time I think we've brushed up against the minimum amount of exercise that can provide substantial health benefits, something comes up that proves me wrong. In late 2014 we published a study that tracked the benefits of the smallest amount of exercise my lab has tested: three 20-second sprints per day, totaling a minute's worth of hard exercise per day amid a total per-day time commitment of 10 minutes. Repeated three times in a 7-day period, the protocol amounted to 3 minutes of hard exercise per week. We asked sedentary, overweight, and obese men and women in their twenties and thirties to follow the protocol for 6 weeks and were astonished at the results. Just 3 minutes of intense exercise per week reduced blood pressure by 6 to 8 percent, and elevated cardiorespiratory fitness by 12 percent, translating into a reduced risk of dying and developing chronic diseases.

Most remarkably, research we've just conducted in my lab tracked the 1-minute-interval protocol's effects over 12 weeks on sedentary, nonathletic individuals and compared the benefits to those of another group that conducted 135 minutes a week of moderate aerobic exercise. And the benefits were the same.

That's right: It was possible for everyday, nonathletic sedentary individuals to derive the cardiorespiratory benefits of 135 minutes a week of traditional endurance training—three 45-minute sessions per week—with just a single minute's worth of hard exercise repeated three times per week.

Who Should Do It? • So far we've only tested the One-Minute Workout on healthy individuals. Assuming you're up for it and have the all-clear from your doctor, the protocol represents the most time-efficient workout that we've tested, to date.

THE WORKOUT

1. Warm up with some light physical activity for 3 minutes at an easy pace.
2. Blast through a 20-second sprint at an all-out pace.
3. Rest with some light activity at intensity 1 for 2 minutes.
4. Blast through another 20-second sprint.
5. Repeat the cycle until you've completed 3 sprints.
6. End with a 2-minute cool-down for a total duration of 10 minutes.
7. Feel free to customize the sprint activity to any full-body movement that significantly elevates your heart rate—such as the stair climbing I mentioned earlier in this chapter.
8. Note that the protocol we tested in the lab featured different warm-up and cool-down times. To bring this workout in line with the others in this book, I've used a 3-minute warm-up and a 2-minute cool-down.

The Tabata Classic

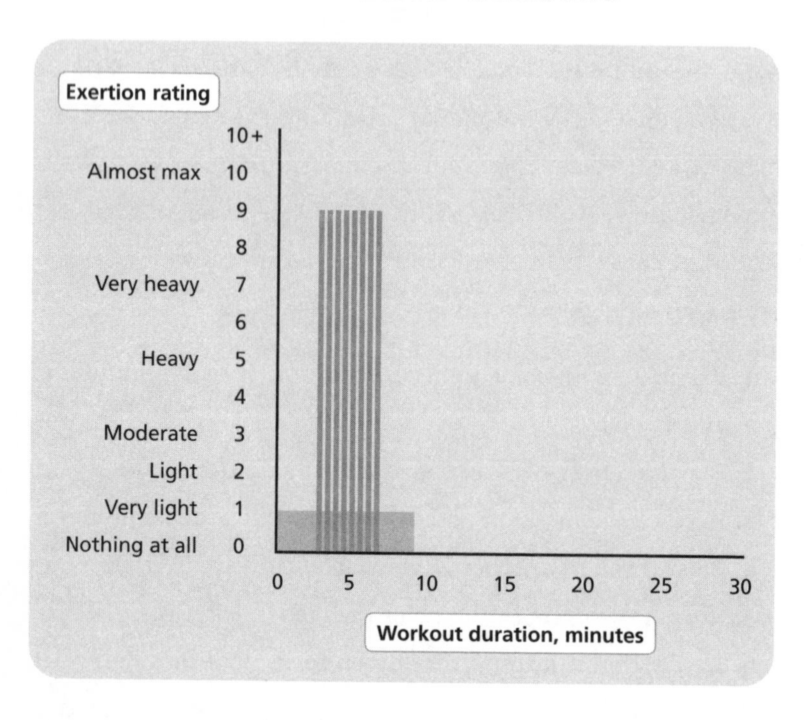

In the 1990s, the head coach of the Japanese speed skating team, Irisawa Koichi, had his athletes employ a brief but intense workout that featured short bursts of high-intensity exercise followed by even shorter rest periods. A coach on the team, Izumi Tabata, was the first to analyze the workout's effect, lending his name to a protocol that helped kick off the interval-training movement. Many personal trainers still love the format today. Note that

Tabata's study used exercise bikes, although virtually any activity that significantly elevates the heart rate can be used.

Peak Intensity • 9

Duration • 9 minutes

The Evidence • Tabata's 1996 paper compared what's now known as the Tabata protocol with moderate-intensity endurance training. The endurance group performed 5 days of endurance testing per week for 6 weeks, while the sprint group conducted the interval protocol four times a week, plus one 30-minute steady-state workout per week at a moderate pace. After 6 weeks, the aerobic capacity of the endurance group hadn't improved at all. But the sprint group, which had conducted hard exercise for a little less than 11 minutes per week, had improved its aerobic capacity by 14.6 percent—making Tabata's study one of the first to show how potent sprints could be as a tool to improve aerobic capacity.

Who Should Do It? • Tabata originally conducted the study with athletic college-age phys ed students, most of whom were on varsity teams for such sports as soccer, basketball, and swimming. With recovery periods that are shorter than the sprint periods, and near-maximal effort, Tabatas are tough. Most people who perform Tabata-style workouts consider themselves hard-core athletes.

THE WORKOUT

1. Warm up at an easy pace for about 3 minutes.
2. Sprint for 20 seconds at intensity 9—not quite all-out.
3. Rest for 10 seconds.
4. Repeat the 20-seconds-on, 10-seconds-off cycle for a total of 8 sprints.
5. Cool down with some light activity for 2 minutes, for a total workout duration of 9 minutes.

A VARIATION

It's easy to include resistance training within the Tabata protocol. In 2012 my colleagues at Queen's University and the University of British Columbia's Okanagan campus created one of the most elegant ways to meld time-efficient intervals with resistance training. They based it on the Tabata format but replaced the sprints with resistance-training exercises—with remarkable results.

Recreationally active female university students conducted the protocol, although the benefits would apply to healthy men and women of all ages. With 4 minutes a day of intense exercise, the subjects improved aerobic fitness as much as a comparison group did conducting vigorous-intensity endurance exercise for 30 minutes per day. In addition, the test subjects increased the number of leg extensions they could do by 40 percent, the number of push-ups by 135 percent, and the number of sit-ups by 64 percent, among other measures. The takeaway? A Tabata *bodyweight* protocol is a potent way to simultaneously boost aerobic capacity and muscle strength. Here's how to do it:

1. Each training day, begin with a 3-minute warm-up of your chosen exercise by performing light versions at a slow pace.
2. On Day 1, complete as many burpees as you can in 20 seconds. To complete a burpee, begin in a standing position. Squat down, place your hands on the floor with your palms down, and kick both legs out behind you so that you end in a plank position. Return to a squat position, and leap up into the air as high as you can while raising both arms above your head. To increase the difficulty, perform a push-up from the plank position.
3. Rest for 10 seconds.
4. Perform the exercise for 20 seconds, trying to complete as many reps as possible. Cycle through the 20-seconds-on, 10-seconds-off format until you've completed 8 bursts of the exercise.
5. Cool down for 2 minutes, for a total workout duration of 9 minutes.
6. On subsequent training days, perform the 20-second-on, 10-second-off cycle for 8 repetitions with the following exercises, concentrating on just one exercise per day: mountain climbers, jumping jacks, or squat-thrusts.
7. To perform a mountain climber, start in the plank position. Staying in the plank, bring one leg forward so the knee

approaches the chest. Return to the regular plank. Then bring the other knee forward so the knee approaches the chest. Repeat.

8. To perform a jumping jack, begin standing up straight with your hands at your sides. Jump and land with your legs set slightly wider than shoulder-width apart and your arms raised above your head so that the fingertips nearly touch. Do a second jump to return to the standing position, with your arms hanging by your sides. Repeat.
9. To perform a squat-and-thrust, begin standing straight with your hands at your sides. Squat down until your hands are flat on the ground. Kick your legs out behind you so that you end in a plank position. Bring in both legs at the same time to return to a squat, and stand up with your arms at your side. Repeat. Note that the study subjects used 5-pound dumbbells, but in the interests of simplicity we've dispensed with the equipment. To increase the difficulty of the squat-and-thrust, incorporate a push-up when planking.
10. Eager for more variety? Rather than doing one exercise per day, consider changing things up so that you're alternating circuit-style between all four exercises within the protocol on the same day.

The One by Four

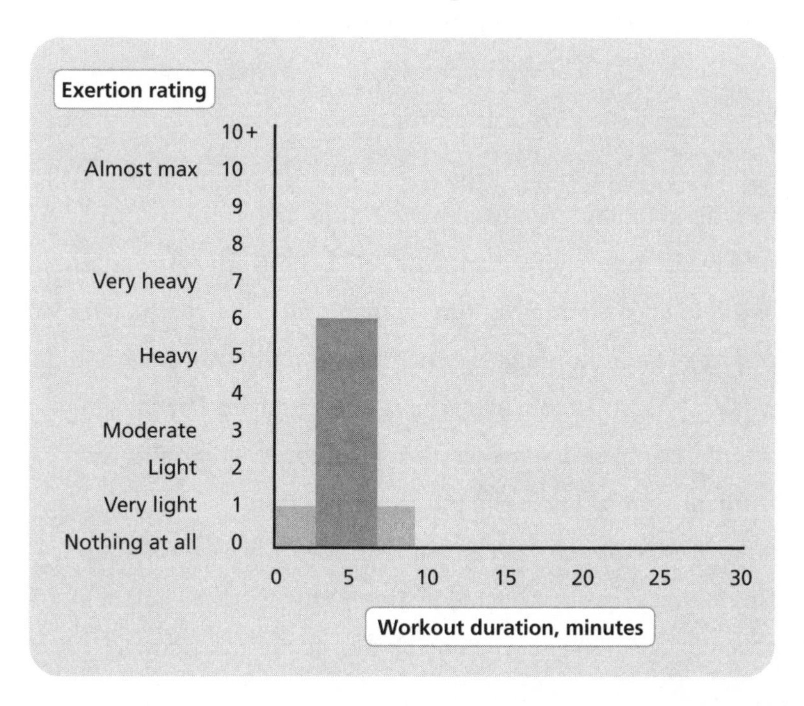

One option for the fit but severely time-constrained is to try to get in at least one good interval. That's the takeaway from a paper I coauthored in 2013 with Ulrik Wisløff, which showed that most of an interval workout's benefits come from the first sprint. Note that Wisløff's lab had its subjects conduct a 10-minute-long warm-up and a 5-minute-long cool-down. In the interests of time efficiency I've cut down both.

Peak Intensity • 6

Duration • 9 minutes

The Evidence • It's possible to reduce the risk of death from cardiovascular disease with a lot less exercise than the guidelines suggest. Ulrik Wisløff and his researchers wondered how little exercise was required to provide a benefit—and discovered something surprising. His lab gathered a number of overweight but otherwise healthy middle-age men and divided them into two groups: one conducted the Norwegian protocol of 4 repeats of 4-minute-long sprints; the other conducted just a single 4-minute-long sprint three times a week. After 10 weeks, Wisløff's lab was surprised to discover that the single-sprint group had experienced most of the benefits of the 4-sprint group. The 4-sprint group boosted its cardiorespiratory fitness by 13 percent, an impressive amount—but the single-sprint group boosted its fitness by nearly as much: by 10 percent. The point? If you don't have much time, a single sprint can pack a powerful punch.

Who Should Do It? • A 4-minute sprint at an intensity of 6 is no joke. If you find yourself unable to do it, simply start with a shorter sprint and work your way up to 4 minutes over time.

THE WORKOUT

1. Warm up with light activity for 3 minutes.
2. Conduct the sprint for 4 minutes at an intensity of 6—fast enough to get you breathing hard but not quite gasping for breath.
3. Cool down with 2 minutes of light activity.

The Go-To Workout

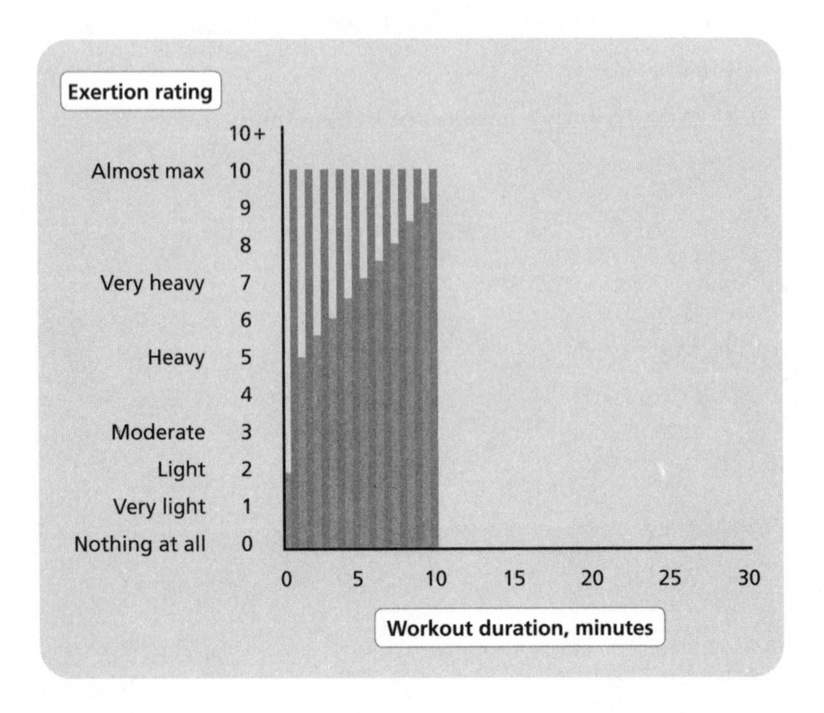

If I could only do one type of workout, it would be this one. My version of a do-anywhere, time-efficient workout is designed to simultaneously enhance strength and cardio-respiratory fitness. It includes some of the best elements of the most time-efficient workouts in this book, including body-weight training for upper- and lower-body strength and active recovery periods that keep the heart rate elevated for cardiovascular training. It's a workout I can do in my basement, where I have a stationary bike and a pull-up

bar. It could easily be adapted to a park, with an open field for running sprints and monkey bars for pull-ups. You could also use a rowing ergometer. Feel free to choose other body-weight exercises, especially if you don't like (or can't perform) a pull-up. Other modifications: You can make the push-ups easier by using your knees, rather than your feet, as the fulcrum. You could also aim to do a set number of repeats instead of blasting through as many as possible during the 30-second intervals. Although this is my go-to, I don't recommend ever sticking to just one workout, because human physiologies and attention spans thrive on variety.

Peak Intensity • 10

Duration • 10 minutes

Who Should Do It? • Any healthy person who feels up to it, basically, and who desires a time-efficient workout that targets the whole body.

THE WORKOUT

1. As a warm-up, perform 30 seconds of jumping jacks.
2. Alternate bodyweight resistance-training exercises with some type of cardiovascular exercise in repeating 30-second intervals. The bodyweight exercises should be performed hard, at an intensity of 10, such that you "fail" or are unable to perform any additional repetitions at the end of the 30-second period. Reduce the intensity somewhat

during the cardio intervals in between, but the pace should remain vigorous, perhaps starting out at an exertion of 5 and progressing to an 8. So while these are "recovery" intervals in between the bodyweight exercises, your heart rate remains high throughout the entire 10-minute workout, providing an effective cardiovascular training stimulus.

3. The bodyweight intervals should incorporate upper- and lower-body exercises. One great combination is push-ups, pull-ups, and air squats. If you're unable to conduct the exercise for the whole 30-second interval, just do as many as you can. Also, feel free to work in such other exercises as mountain climbers, burpees, or lunges.
4. The cardiovascular exercise could be cycling, climbing stairs, or running a predetermined "lap" around a park or even briskly in place. You could stick with one type of exercise or vary this throughout the workout.
5. And you're done! Congratulations—you've just employed a variety of the most potent, scientifically proven fitness- and strength-boosting techniques to improve health, in only 10 minutes!

CHAPTER EIGHT

High-Intensity Nutrition

MOST OF THIS BOOK HAS BEEN ABOUT HOW TO GET FIT, fast. To get healthy, which sets you up to live a long and active life. However, human beings tend to care about more than just being healthy. We want to feel good, yep, definitely. But we also want to *look* as good as we feel.

Exercise does a lot of things. It increases your fitness, improves your health, reduces your risk for chronic diseases, and extends your life expectancy. In my opinion, interval training is the most potent form of exercise—it's the most powerful way that physical activity can get you healthy.

But if you want to employ the most efficient levers to improve your body composition—lose fat, gain muscle mass, look better—then you also need to address your diet and nutrition. You need to change what goes into your mouth.

So that's what this chapter's about. Many other books out there will tell you much more about healthy diets. You want to become an *expert* in all things diet-related? Want to pursue a career as a nutritionist or dietician? This is not the chapter you seek. Most of us don't need or want to become nutrition experts. Consider this chapter a quick-and-dirty approach. Once you've read it, you'll know enough about how to lose weight to get it done, plus some little tricks to help you eat a better diet—you'll get the essentials here to assist your effort to go from pudgy to chiseled in the most efficient manner possible.

Fitness and Fatness

Before we get to your looking better, I want to give you a potential pass here. Because from a pure *quantity*-of-life perspective, there's an interesting relationship between fitness and fatness—one that most people find surprising. Lots of research suggests it's more important to be fit regardless of whether or not you weigh too much. A prominent name in this field is Steven Blair, an exercise science professor at the University of South Carolina who has conducted many large studies that look at the intersection of fitness, fatness, and mortality.

In general, his studies show what most of us expect—that in

the continuum of people, risks for all sorts of health conditions increase with one's weight.

However, as Blair and his fellow researchers dug into their data, they noticed something interesting. The mortality risks of overweight people changed radically if they were *fit*. It may seem shocking, but they discovered that overweight people *who are fit* have mortality risks similar to the risks of people of normal weight. "In all of these studies, we typically see higher rates of mortality, chronic diseases, heart attacks and the like, in people with high BMI [body-mass index]—we see the same thing everybody else sees," Blair told the *Guardian* newspaper. "But when we look at these mortality rates in fat people who are fit, we see that the harmful effect of fat just disappears.

"If we look at individuals who are obese and just moderately fit . . . their death rate during the next decade is half that of the normal weight people who are unfit. So it's a huge effect."

The message is, if you're carrying around a few extra pounds, or even more than a few, and you want to improve the chances that you'll live a longer life—you don't necessarily have to lose weight. It's more important to start exercising. Ideally it's best to do both. But you really need to get fit.

Maybe you start with some interval walking. Progress to the Ten by One or the Fat Burner. Eventually, you're able to hit the stairs at lunchtime and slam through the One-Minute Workout.

Ideally, you should find ways to incorporate more physical activity into your daily life. Maybe you decide to commute with a bike rather than a car. Maybe after work you go out with your kids to play some touch football. Or meet up with your friends to take in a spin class *before* work. The point is, you get *active*. Then, chances are you're going to live just as long as fit people of normal weight. Chances are, you'll live longer than those of normal weight who *aren't* fit. And the chance that you'll develop chronic illnesses like type 2 diabetes or heart disease substantially decreases.

The takeaway can be summed up neatly: Being fat but fit is better than being normal weight (or even very slim) but unfit. So think about getting comfortable with your body and consider pursuing fitness instead. As Blair and his colleagues concluded, "These findings are promising for all individuals, including those unable to lose weight or maintain weight loss, as all can experience significant health benefits (by) participating regularly in physical activity."

Interval Training and Body Composition

Still, most of us would prefer to be of normal weight, regardless of how fit we are. Being overweight can infringe on what we do and where we go in all sorts of ways. We stay at a normal weight, and we're more likely to engage in the routine activities that people tend to do everyday without needing assistance. We tend to have more energy. We're more likely to seek out new experiences and to fully participate in all the opportunities offered by this wonderful world of ours.

We'll look better, too.

So let's assume that if you're carrying around a few extra pounds of fat, or more than a few, that you're interested in shedding those.

The human body provides us with two main methods that can help achieve that goal of cutting fat: exercise and diet. By far the most efficient method is diet—and we'll discuss some great tips on how to pursue that method in a moment. But exercise *can* help you change your body composition, too.

One of the earliest studies to examine the effect of intervals on body composition, published in 1994, was conducted by a team out of Quebec's Laval University. It was an ambitious study that compared intervals with traditional moderate-intensity training over the course of twenty weeks. Neither group lost weight overall, but there were some changes in body composition. As measured by skinfold thickness, the endurance-training group cut their fat at three of the six points on the body. The interval-training group cut fat everywhere the researchers measured—the back and front of the upper arm, the lower leg, the abdomen, the hip, and the upper shoulder. Plus, the interval trainers lost more fat, as measured by skinfold measurement. "For a given energy expenditure of activity," the researchers concluded, "fat loss is greater when exercise intensity is high."

A more recent study published out of the University of New

South Wales in 2008 followed forty-five young women over fifteen weeks. A third of the subjects performed twenty minutes of intervals. Another third performed forty minutes' worth of steady-state exercise. The final third, a control group, did no training of any sort.

I've mentioned the results of this study in chapter six, where this protocol is listed as the Fat Burner. The protocol format saw the interval subjects mounting a bicycle and going as hard as they could for eight seconds at a time, then resting for twelve seconds, until they'd reached a total of sixty repetitions over the course of twenty minutes. In contrast, the steady-state exercisers simply cycled at 60 percent of their VO_{2peak} for as long as they could, until a maximum of forty minutes. The scientists calibrated the study so that the two groups were expending similar amounts of energy throughout their exercise sessions.

After the fifteen-week intervention, the total bodyweight, fat mass, and abdominal fat decreased more for the interval trainers than it did for the endurance trainers. That's a remarkable result, given that the interval trainers exercised for just 36 minutes per week (excluding warm-up and cool-down) while the endurance trainers exercised for 120 minutes per week. "Despite exercising half the time, [interval training] subjects in the present study lost 11.2 percent of total [fat mass] with [steady-state exercise] subjects experiencing no fat loss," the researchers noted.

Which Burns More Calories—Interval or Endurance Training?

Lots of personal trainers will tell you that HIIT burns fat like no other workout. And that's true. Others will criticize HIIT, saying its short durations don't burn as many calories as longer-duration moderate exercise. That's *sort of* true.

How can both things be right? How does HIIT burn more fat *without* burning more calories during the exercise?

I'll explain.

Those who exercise at moderate intensities for long periods of time tend to burn more calories *during* the period they're actually exercising, as compared with the amount burned by those conducting an interval workout. Things change once the workout's over.

Exercise elevates metabolism, or the rate at which the body uses oxygen to burn fuels for energy. The more intense the exercise, the greater the effect on metabolism—and the longer it takes for the body to return to its normal resting state. Technically speaking, that period of elevated metabolism during recovery is called *excess post-exercise oxygen consumption.* Many personal trainers refer to it as the *afterburn*. The more intense the exercise, the greater the number of calories expended during the afterburn. And the afterburn's calorie-burning effects are greater after intense exercise than after moderate exercise.

For example, the Ten by One interval protocol might burn half the calories that get used during fifty minutes of continuous moderate exercise—even though the two workouts would have similar fitness effects in terms of boosting VO_{2max}. But then, once you've stopped exercising, the interval trainer's body will continue to burn extra calories at a greater rate than the moderate exerciser's.

Researchers in my lab were curious about how many calories were consumed during this afterburn. So we set up an experiment that measured subjects' calorie burning over a twenty-four-hour period, starting with an exercise session and then for the remainder of the day and overnight period. We had a group of subjects conduct our Ten by One interval protocol one day and on a separate day conduct fifty minutes of continuous exercise at a moderate pace of 70 percent of maximum heart rate. Then we calculated the amount of calories both groups burned over a twenty-four-hour period, including what they burned during their workouts.

What we found was pretty amazing. Remember that the continuous exercise lasted more than twice as long as the HIIT and required our subjects to do twice as much work. And yet, over the course of a twenty-four-hour period, the subjects burned similar amounts of calories regardless of which type of workout they conducted. The effect of the intense exercise's afterburn was such that, regardless of whether they engaged in continuous or interval-based exercise, they ended up burning the same amount of calories over the twenty-four-hour period in which they conducted the exercise. And other labs have demonstrated the same effect.

So while you may not burn as many calories *during* the

interval-training workout, its afterburn effect is such that it burns a similar amount of calories as a continuous-exercise workout over the long term.

How to Lose Weight

As you can probably tell by now, I'm all about efficiency. I don't want to waste time doing things in a certain way when there's a more time-efficient way to do something. And the most effective and efficient way to lose weight is to reduce the amount of food that you put into your mouth. Not an earth-shattering concept, right? Weight gain or loss is mainly about how much energy you put into your mouth. Eat more calories than you burn, and you gain weight. Eat less, and you lose it.

That's simple enough to say. Most people *know* that. And yet two out of every three Americans are overweight, and one out of three actually is obese. Most of these people would *like* to lose weight, in theory.

And yet they don't.

So, clearly, the reasons for why so many people are overweight and obese are complex. Consider the following as some tips that might help.

First, some basic things. Then after that I'll provide five tactics designed to help you drop the pounds—or at least prevent you from gaining more. The first thing to do is figure out how many calories you need to eat per day to maintain your weight: a figure that amounts to one's metabolic rate, most often expressed as the average number of calories you burn a day. Some health clinics and exercise laboratories will charge you a lot of money to identify your metabolic rate. The most accurate way to measure the number of calories you expend in a day uses a technique called direct calorimetry, in which scientists analyze the body's heat production. The subject goes into a little chamber equipped with sensitive thermometers that measure the total amount of heat the body gives off. That value can be used to determine the number of calories required by the body.

Another technique is indirect calorimetry. This is what we use in my lab. The subject sits at rest and breathes through a tube so we can measure the amount of oxygen consumed and carbon dioxide produced. It's essentially the resting part of a VO_{2max} test, but in a more controlled environment. We do that for a certain fixed amount of time, like a half hour. The gas exchange measurements provide a very accurate estimate of the number of calories burned per day.

But you can skip these super-accurate techniques unless you really want a precise individualized test. All you probably need is a ballpark measurement of the number of calories you burn. A reasonable estimate.

To get that, you can just use an online calculator. Plug "metabolic rate" into your favorite search engine and then click around until you find one that uses your age, height, and weight. It should also ask you to estimate your activity level.

Then the calculator will give you a number—say it's 2,800 calories a day. Bingo—you have your estimate of the number of calories you need to maintain your weight.

Would it be more accurate to have done this by direct calorimetry? Yes, but this is good enough for our purposes.

Next, determine how many calories you need to remove from your diet to lose weight in a healthful and steady manner. This all depends on how quickly you *want* to lose weight. You could cut your caloric intake in half if you wanted, and you'd lose a lot of weight in a comparatively short amount of time, but you also might feel like hell while you're doing it.

To learn about this next part, I turned to my friend Stuart Phillips, who is a professor in my department and the director of the McMaster Physical Activity Centre of Excellence. We've been colleagues for a long time and were even hired on the same day. He's conducted numerous studies on people who are trying to go from overweight and underfit to lean and in shape. As a suggestion of the number of calories you should cut from your daily average, Phillips's number is 500—as in, cutting 500 calories from one's daily caloric intake—which means a drop of between 15 and 20 percent for some people. So a person who needs to consume 2,800 calories to maintain weight would cut the daily caloric intake to 2,300 calories. And for your reference? A single Big Mac from McDonald's has about 540 calories.

A general rule of thumb says cutting 3,500 calories from a diet over time will lead to a drop in weight of a single pound. So cutting 500 calories per day puts you on track to lose about a pound a week in weight. (As in most things diet-related, this is an estimate, which you're not likely to meet exactly because of individual differences in metabolism.)

You can cut more, just like you can cut less. To some people, a pound a week doesn't sound like much—they want a pound *a day*. Others are suspicious of fast weight loss. They may have heard the much-discussed aphorism that those who lose weight quickly will gain it back quickly, too. That slow weight loss puts you on track to stay lean longer.

Phillips says that's not actually true. Citing long-term academic studies, he notes the speed at which one loses weight has nothing to do with how long one is able to maintain a healthier weight. In fact, it's possible that *rapid* weight loss is better, because the people who lose weight quickly tend to be more motivated than those who take longer to shed the pounds.

Another thing that doesn't matter much in terms of the overall success of the weight loss is the precise kind of diet that one pursues. "For every diet that's out there you can find someone who's supported the approach," Phillips said. "If you want to go all carbohydrates, high fiber, and low fat, that's the Dean Ornish diet. Or if you want to go all the way to the other side, and have almost no carbs but lots of protein and fat, then that's Atkins. It's really confusing. You can lose a lot of weight on either one—so long as you stick with it."

So find a diet that fits your sensibility and that you believe in. It doesn't really matter which one. Next, you're going to need something to help you count your calories. A lot of the subjects in Phillips's studies use the app MyFitnessPal, which allows you to keep a running tally of your energy consumption throughout the day.

One final thing before we get to our five tactics to help speed the weight loss. If there's a fundamental truism, Phillips says, it's that the dieting person should try to eat as many meals as possible that consist of actual food: fruits, vegetables, meats, and dairy. Avoid processed foods with lots of different ingredients. It's better to prepare your food yourself, because who knows what the restaurants are putting into the meals. And this might sound like common sense, but avoid foods that have a lot of calories and few nutritional benefits, such as potato chips or any form of candy.

Now, here are Stu Phillips's five tips for losing weight in an efficient and a sustainable manner.

1. Eat Protein with Every Meal

Protein is needed by the body for lots of things, including building muscle. You need protein if you want to grow stronger and if you want to improve your appearance by becoming more muscular. But there are a lot of other reasons to eat protein as well.

For one, protein dictates hunger levels. You eat a little protein during a meal and you're more likely to feel full. You eat *more* protein, and you feel *more* full. Which staves off your feelings of hunger longer. That's important to someone who is trying to lose weight, because the more full we feel, the better our choices when it comes to food.

When losing weight, the average person tends to lose three parts of fat to one part muscle tissue. Faster weight loss shifts that ratio even more toward muscle loss.

Obviously, as Phillips points out in a recent review in the

academic journal *Sports Medicine*, it would be preferable for both human appearance and function to maximize the fat loss and minimize the loss of muscle. And the best situation for the person who wants to improve body composition quickly would be to lose fat while *gaining* muscle.

If you're trying to change your body composition, you should be consuming protein at every meal, Phillips says, and as close to the same amount each time. See, at breakfast, thanks to some peanut butter on toast, or maybe some eggs, most people get about 10 to 12 grams of protein. Then thanks to the turkey or roast chicken in their sandwiches, maybe they get 15 grams at lunch. Dinnertime comes and they have a chicken breast or a pork chop or maybe even a nice steak. At dinnertime, many people consume around 70 or 80 grams of protein.

Not a very even pattern, right? A little protein, another modest serving, then a *lot*, and then nothing until the next morning—that's not the optimal way to consume protein if you're seeking to efficiently prevent hunger or to provide your body with the raw material it requires to make and hold on to lean muscle tissue.

Think about the times when you're most likely to make poor food choices. For many people it's mid-morning and mid-afternoon. Mid-morning, it's heading down to the coffee shop for a sugar-laden, caffeinated pick-me-up that might have 500 calories or more. Mid-afternoon, people may head down to the vending machines for a chocolate bar. Not a good food choice.

So the snacking happens a couple of hours after breakfast and lunch, the two meals where we tend to consume the least amount of protein. Notice a pattern?

To even out hunger pangs, Phillips suggests getting about the same amount of protein each time you eat. "I make sure at breakfast that I get a pretty substantial serving of protein so that I don't feel the need or the temptation to snack midmorning," Phillips says. "The product I eat a ton of these days is Greek-style yogurt." (Which has about 16 grams of protein per three-quarter cup serving.)

He also eats a lot of eggs, which were taboo for years but now are being reconsidered as a healthy food choice as research is showing a disconnect between consumed cholesterol and blood cholesterol. "So eggs have undergone a bit of a renaissance—they're back on the menu," Phillips says.

"We can get by without fats and without carbs," he says. "But we can't get by without protein—it's the only macronutrient for which we have an absolute requirement. That's why bodybuilders will set their alarms and wake up in the middle of the night to swallow some protein. It's to ensure the body has enough to build muscle."

I'm not saying you have to set your alarm to eat in the middle of the night—that seems a bit drastic. But if you're trying to lose fat while retaining or even building your lean muscle mass, as many interval trainers are, then it makes sense to ensure your muscles have the fuel stores they require.

Phillips conducted the research that established the optimal dose of protein required to retain muscle mass while losing weight. Remember that 3:1 ratio of weight loss, where you lose a pound of muscle for every three pounds of fat? Phillips has found that you can shift that ratio by eating a lot of protein. To reduce muscle loss while losing weight, he suggests eating at least 0.25 grams of protein per kilogram of bodyweight at every meal (which works out to 0.11 grams of protein per pound of bodyweight). So the optimum per-meal dose for a 130-pound woman who wants to maximize her ability to retain, or perhaps even build, her lean muscle mass would be about 15 grams—or what's in that three-quarter-cup serving of Greek yogurt. The ideal protein serving for a 190-pound man would be a little more than 20 grams of protein. (If you want to *gain* muscle while losing weight, Phillips suggests upping your protein intake even more, to 0.22 grams per pound of bodyweight.)

When saying "per meal," I don't mean the 7 a.m., noon, and 6 p.m. frequencies of breakfast, lunch, and dinner. Rather, academic studies have shown the muscles like it best when they get the optimum dose of protein every four hours.

Consequently, Phillips suggests that interval trainers looking to maximize fat loss while also minimally losing muscle mass eat their optimal 0.11 gram per pound of bodyweight every four hours throughout the day with a double dose taken immediately before bed. That amounts to one dose of protein at 6 a.m., more at 10 a.m., 2 p.m., and 6 p.m., and a double dose at 10 p.m. Here's hoping that the vegetarians among us like eating tofu, nuts, and whey protein powder. For all the rest of us—enjoy the chicken!

2. Don't Drink Your Calories

One of Phillips's most effective tactics for weight loss also happens to be one of the easiest: Drink water. That is, aside maybe from the occasional coffee, beer, or glass of wine, don't drink anything *but* water. How much water? That depends on the person—most of us get along fine by simply using thirst as a guide.

Cutting sugar-laden drinks like soda and juice from your diet is likely to be a big change if you're anything like the average American. There's a tremendous amount of information available on soda consumption, because governments and other agencies became alarmed about the issue and the way it was affecting obesity rates. From the 1960s to 2001, soda consumption in America increased dramatically, until the United States was the world's largest consumer of carbonated soft drinks. Near the peak rate of soda consumption, in 2001, sugary drinks formed about 9 percent of Americans' daily caloric intake, up from 4 percent in the 1970s, according to a Harvard School of Public Health sugary-drink fact sheet. The average American adult drank 170 liters of soda annually in 2011. That is a lot of soft drinks. But in recent years, as the public has grasped the extent that soda calories have contributed to the obesity epidemic, Americans have moved away from drinking soda and are instead choosing water. Argentina surpassed the United States in soda consumption in 2014, and according to an excellent 2015 *New York Times* feature titled "The Decline of 'Big Soda,'" one industry consultant projects that by 2017, bottled water sales will become the single largest US beverage category for the first time, surpassing soda. Between 2004 and 2012, according to the *Times*, children consumed 79 fewer calories per day from sugar-sweetened beverages, the single largest product-category shift.

The problem isn't only how much sugar is in soda—though it's a lot. One 20-ounce bottle of soda, for example, has 19 teaspoons of sugar and 290 calories. Another problem is that, despite all the calories, the soda doesn't make one feel full. It's possible to have a couple of Cokes, quaffing hundreds of calories, and *still* crave a burger and fries or a bag of chips.

Fruit juices can be nearly as bad as sugar-sweetened beverages. For example, orange juice contains more calories than do cola-based sodas—about 112 calories per cup of OJ versus 97 calories per cup of Coca-Cola. The Coke has about 7 teaspoons of sugar per cup; the OJ 6 teaspoons. Apple juice has similar amounts of calories and sugar as orange juice.

What about sports drinks? Gatorade has roughly half the sugar and calories of Coke, while the brand's more calorie-conscious G2 line of sports drinks has about a fifth of the sugar and calories.

All those calories add up. So if you're trying to lose weight, one easy way to do it is to eschew all the liquid calories you can. Which means, drink water. And that's it. Instead of sports drinks and juice. *Definitely* instead of soda. Just drink water.

Finally, avoid resorting to the diet versions of sodas, which are loaded with artificial sweeteners. An Israeli team of researchers published a 2014 study in *Nature* suggesting that compounds like aspartame alter the makeup of bacteria in the gut in such a manner that it encourages glucose intolerance. The situation could lead to a greater risk of developing diabetes and, in a stroke of irony, may encourage obesity.

You can't be perfect all the time. Even Phillips allows himself to reach for a beer now and then when he feels like he's had a good week and deserves a reward. And that's great. Rewarding yourself for good practices reinforces your likelihood of making a good choice the next time around. On all other occasions, keep your beverages clear: Drink water.

3. If You Must Snack, Choose Almonds

It's late at night, the kids are asleep, and you're settling down with a book or tablet computer, catching up on social media, maybe viewing an episode of whatever show you happen to be into. And the impulse strikes—it'd be great to have a snack right now. I get it, too. One of my weaknesses happens to be Lay's plain potato chips. I can tear open a bag and do some real damage to it. On occasion, I've been known to eat an entire bag in one sitting.

The problem is that chips are Phillips's ultimate bugaboo.

They're energy dense but nutrient light—a prototypical junk food. An ounce of plain potato chips has 160 calories, but no one stops eating after a single ounce. One of those large 9-ounce bags has a remarkable 1,440 calories. That's about half the calories I'm supposed to get in a day. And a whole lot more salt than I need.

Rather than chips, Phillips suggests reaching for nuts when you're craving a snack. People used to disparage nuts because they're full of fat. And they are, but it happens to be heart-healthy polyunsaturated fat. Phillips's preferred nut is the almond, which has experienced a bit of a brand makeover in recent years. "A decade ago, almonds were a snack," says Phillips. "Now they're a health food. They've completely changed in the eyes of most health professionals."

Almonds' caloric content is about the same as chips'—about 160 calories per ounce. But there are a couple of bonuses with the nuts. At 6 grams of protein per ounce of almonds, they're packed with the stuff that fills you up. That same ounce has 3.5 grams of fiber. They also have a lot of other nutrients, such as vitamin E, manganese, magnesium, and copper. Most times, a handful will stave off the hunger pains.

Another thing about nuts, Phillips says, is that they're comparatively hard to digest. It's difficult for the human body to extract all the calories they contain. The food energy in chips is relatively easy for the body to whisk into fat stores. But the rigid texture of nuts makes them difficult to chew, which in turn prevents the body from efficiently extracting all the food energy they contain. So while you might *swallow* 160 calories of food energy in an ounce of nuts, a lot of that actually is getting excreted out of you sometime later.

Finally, Phillips says nuts are negatively correlated with all kinds of disease. The more nuts you take in, the lower your risk for cardiovascular disease, diabetes—you name it. "Almonds," he says, "are my go-to snack food."

4. Fast Intermittently

We're mentioning this technique because we're trying to help you go from out-of-shape and chubby to fit and lean in the most efficient manner possible. You don't have much time. You want your changes to happen *now*.

If that's the case, then you'll be interested in one of the most fascinating approaches to weight loss to hit the academic literature in years: intermittent fasting. Starving yourself. Not eating at all for a certain amount of time at regular intervals. Voluntarily.

Humans have fasted for all sorts of reasons through the ages. There's the Jewish practice of fasting on Yom Kippur and the Muslim practice of refraining from eating during daylight hours during the month of Ramadan.

Science has accumulated considerable knowledge about fasting's effects on the human body. It's actually possible to fast for a lot longer than you might expect. Alex Johnstone of the University of Aberdeen in Scotland has written an excellent overview on the topic of fasting. According to Johnstone, the longest recorded fast anyone survived began in 1965, when a twenty-seven-year-old man who weighed 456 pounds chose to attempt to lose weight simply by not eating. The fast was supervised by doctors from the University of Dundee in Scotland. They provided him with potassium tablets and multivitamins, and he ended up losing 275 pounds, or about 60 percent of his bodyweight. The duration of the fast? More than a year, at 382 days.

The point? Short-term fasting of a day or two won't do you any physical harm—although you should be drinking water throughout that time. In fact, Phillips says, we have data on many different animals that suggests restricting food leads to longer lifespans. That is, tests conducted on all sorts of animals have shown that extreme calorie restriction on the order of a 20 to 40 percent decrease from what they would normally eat results in all sorts of beneficial changes, including longer lifespans, a decreased cancer risk, and improved insulin sensitivity.

Do humans benefit from fasting? Those long-term studies haven't been done—but there is a community of people in North America who believe it's healthy to subsist on an extremely low-calorie diet. The movement is known as calorie restriction, and it originated with a UCLA medical school professor named Roy Walford. As one of eight crew members in an experiment in isolated, self-contained communal living called Biosphere 2, which lasted from 1991 to 1993, Walford convinced his fellow bionauts to attempt to live according to the low-calorie, high-nutrient diet that calorie restriction required. When the bionauts left their facility, they'd substantially improved their blood cholesterol, blood pressure, and other health markers.

Today, a calorie-restriction subculture thrives on the Internet. It's full of people who get by on daily energy consumption levels from 1,300 to 2,000 calories, depending on their weight. The long-term subsistence level leaves them with body-mass indexes of 15 or 16—one six-foot-tall male practitioner weighs just 115 pounds, which is underweight by anyone's measure. (The minimum range for normal-weight individuals who stand six feet tall is 140 to 180—for a BMI of 19 to 24.) And yet, members of the movement insist that their blood markers are the picture of health.

Neither I nor Stu Phillips is suggesting long-term calorie restriction. Rather we're suggesting that people seeking strategies to help them lose weight quickly consider *intermittent* fasting.

Several different approaches exist. One version prescribes fasting one day, then eating normally, as one's appetite dictates, the next. A slightly different version has you eating normally one day and then restricting caloric intake to 25 to 50 percent of normal the next. Another protocol, pioneered by Britain's Michael Mosley in his 2013 book *The Fast Diet*, suggests eating normally for five days and then radically restricting calorie intake the next two, to 500 calories a day for women and 600 for men. As for Phillips, he suggests a much easier form, which sees eating reasonably for six days and then fasting one day of the week.

"Even if you do it only now and again," Phillips says, "intermittent fasting would have a marked effect on your ability to regulate blood glucose, lipids, and other biomarkers of health."

That's because intermittent fasting appears to provide some of the same benefits of calorie restriction—with much less discomfort involved.

For example Alex Johnstone examined what happened to twenty-four male and female participants who fasted for thirty-six hours, from 8 p.m. one night through the whole of the following day until an 8 a.m. breakfast the morning of the third day. Everyone lost weight during the fast—an average of 1.33 kilograms for men and about 1 kilogram for women. What happened *after* the fast helped explain the fasting technique's growing popularity. The day after the fast, the subjects ate more than usual, as expected. But they didn't eat enough to gain back all the weight they'd lost, increasing their caloric intake by only 20 percent compared with the pre-fast day. One mechanism to suggest why this happened is that the subjects naturally ate more fat at their first breakfast meal following the fast—and Johnstone theorized that the feeling of fullness that followed the breakfast "swamped any further increase in the urge to eat for the remainder of the day."

That was short-term fasting, however. What happened when the alternate-day fasting continued? According to Johnstone, one study that saw obese subjects pursue alternate-day fasting for between eight and twelve weeks saw the subjects lose 5 to 6 percent of their bodyweight, with reductions in waist circumference between two and two-and-three-quarter inches. The interesting thing about the study, Johnstone notes, is that obese subjects who followed an alternate-day fast tended to lose their feelings of hunger on the fasting day over time. The end effect, then, is an ability to restrict calories without the concomitant feeling of hunger that usually entails.

Should you be conducting interval-training sessions during your fasts if you choose to pursue intermittent fasting? That depends on you. Some people will get dizzy when they exercise while they're fasting; others will be fine. The real message is, if you feel like you can exercise while you're fasting, go for it. But if you're the type who needs to eat something before you exercise, then you should definitely eat something before you exercise. In other words, use common sense!

5. Exercise

Successful weight loss in general means making small changes on a daily basis that build up over the long term into a substantial impact. It's difficult to be more specific than that. Those who engage in successful weight loss, and keep it off, engage in all sorts of strategies to cut and maintain a decreased caloric intake.

Relying on tactics from the National Weight Control Registry, a database of information about people who have lost at least thirty pounds and kept it off for a year or more, Phillips points out that there doesn't seem to be a single "magic bullet" diet for those who successfully lose weight over the long term. Whether you're doing Paleo, Atkins, the South Beach, or low-carb, high-fat, or whatever, the primary driver of long-term weight loss remains adherence. So long as you stick with any diet, you're going to lose weight.

The National Weight Control Registry does reveal some commonalities among the people who tend to keep off the pounds. Most report that once they've lost the weight, they keep it off with a combination of a low-fat, low-calorie diet and lots of exercise. Roughly 78 percent eat breakfast every day—although, as mentioned above, there's no stat on how many have a high-protein breakfast. Interestingly, 62 percent say they watch less than ten hours of TV per week. And finally, 75 percent of them weigh themselves at least once a week.

But the biggest commonality among the people who have successfully maintained weight loss over the long term is exercise. For example, 94 percent of the people who lost weight and kept it off increased their physical activity, with most of them pursuing walking. And 90 percent of the subjects exercised every day.

Bear in mind that these people aren't necessarily exercising precisely because it burns calories. The reasons they're doing it, and the reasons it helps them to lose weight, are more complex than that.

Here's why exercise as a standalone weight loss strategy is a tough way to shed pounds. Recall the 3,500-calorie rule I mentioned earlier in this chapter? Well, it breaks down after time. Let's say someone's ready to begin losing weight. So they start walking. Walking an extra mile a day correlates with an energy expenditure of about an extra 100 calories per day. Tallied up over a period of five years, the deficit of 100 calories per day should translate into a loss of fifty pounds. In fact, though, a person who begins walking an additional mile every day, all other things being equal, will lose only ten pounds over five years. The complicating factor here is the extent the body changes in response to the walking-caused weight loss. The less body mass you have, the fewer calories you require.

To boil the point down to its most basic element: Let's say you're considering eating a donut, and you're justifying the extra calories by telling yourself that later you'll go for a run to burn off the extra calories you ingested. In my opinion, it's much better to not eat the donut, period. If you're looking to limit your calories, the best strategy is to avoid putting the food into your mouth in the first place.

So why is it that exercise is so crucial to successful weight loss? Physical activity has all sorts of other benefits that directly help keep you on track toward your weight-loss goals. Which is likely why the National Weight Control Registry dieters employ it. For example, exercise boosts mood, fights depression, and decreases anxiety. It acts as a course corrector that functions to keep the dieter on track and focused on life improvement.

So what's more important—diet or exercise? It's a question physiologists get asked all the time. Phillips likes to say that muscles are made in the gym and fat is lost in the kitchen. But the fact is, if you're intent on looking and feeling good, then exercise and a healthy diet *together* are better than either in isolation. Both Phillips and I agree: Do both. "The two together are not just additive, in terms of overall effect," he says. "They're synergistic."

Some Final Food for Thought

I want to leave you with a story. It's about a guy I knew when I was going to the University of Windsor, which is in my hometown, just across the Canadian border from Detroit. At that point I was really into triathlons. For the swimming portion of my training I enrolled in a master's swim program to try to get better. And I got to know a guy named Greg, who was a remarkable athlete. One of the fittest guys I've ever seen. He was about six feet six inches tall and weighed about 170 pounds. In his thirties, he could burn through a marathon in under two hours and thirty minutes. Just to give you some reference, only about 1 percent of people are able to finish a marathon in under three hours. Greg was among the best long-distance runners in the area and even finished in the top 10 in a local marathon one year.

One day at the pool he told me how he became an athlete. He had started out life as just a regular guy. Not really into sports. In his early twenties he got a factory job. He got married, put on some pounds, and was headed toward spending the rest of his life as a couch potato.

Then his wife had a girl's night out to celebrate some occasion and ended up at a male strip club. She came home talking about the guys she'd seen and the fantastic shape they were in. It irritated Greg so much that he said, "You know what? I'm going to get in shape, too." He went on a diet and started running. The pounds just dropped off him, and pretty soon he was a top local triathlete. By this point he was in his thirties, and he was discovering for the first time that he'd been gifted with this freakish physiology, with a VO_{2max} that was off the charts.

The message is, lots of people can be like Greg. Maybe not to the same degree, but we definitely see it in the lab. We'll take a woman who has never really exercised before, hook her up to a metabolic cart, and get her on the exercise bike, and she'll blow a VO_{2max} that's in the top 5 percent for her age group. All that genetic potential just sitting there—and yet she never knew it was there because she'd never challenged herself.

So, yes, some of this stuff is genetic. Some people just happen to have a tremendously efficient engine. Some lose weight quickly. Others lose weight slowly—but they're also apt to hang on to their muscle mass. Phillips found this out in a pair of studies he conducted recently. On young men and women, he's done tests that use the twin tools of fitness and food intake to try to get them as fit and svelte as possible in the shortest amount of time. The results were impressive for some people, and not so impressive for others. Some whipped themselves into shape, and others took longer.

When it comes right down to it, most people can lose weight. It's the keeping it off that is the real trick. Genes play a part. *Lots* of stuff plays a part. So go easy on yourself. Don't be too rigid with your eating. Consider the above rules more as guidelines. Try to adhere to them, and if you don't, well—keep trying. The point isn't to be as strict as possible; it's to be in this for the long haul. To eat healthfully, in reasonable amounts, for years.

Once you get down to around your target weight, you have the freedom to adopt one edict for living that I'm hoping you'll try to follow forever: the 80/20 rule. Sidle away from whatever strict eating regimen got you down to where you wanted to go. An extraordinarily rigid eating plan might set you up for short-term success, but it's certain to lead to long-term failure. So from here on out, 80 percent of the time you should try to eat something good, and 20 percent of the time you should enjoy yourself. That's the reward of being physically fit and eating healthfully most of the time. You can allow yourself the freedom of not worrying too much about the food you're eating.

In the next chapter, about the future of exercise and interval training, we'll further examine the idea that each individual responds differently to exercise, and discuss the evolving science behind providing exactly the right exercise prescription to fit your biology.

CHAPTER NINE

The Perfect Exercise for You

THANKS TO THE GLORY OF NETFLIX, MY WIFE HAS BEEN working her way through the 1990s hit sitcom *Friends*. I caught an episode with her the other day—"The One Where Phoebe Runs," which chronicles the free-spirited roommate's attempt to take up jogging. After going for a run with Phoebe in Central Park, Rachel complains that her new jogging partner runs like "a cross between Kermit the Frog and the Six Million Dollar Man." The storyline's dilemma involves Rachel's embarrassment. She feels self-conscious when she runs in public with Phoebe, and so she fakes a sprained ankle to get out of it.

I think of this episode whenever I hear stories about people conducting intervals out in the open—on a city street, for example, through the course of a regular day. Rather than dealing with the hassle of parking a car, I know people who will take their bikes to work, and they'll punctuate their commutes with thirty-second bouts of pedal mashing that leave their thighs burning and their moods exhilarated. Another friend will bolt for blocks as he goes around downtown Toronto, conducting his errands. He'll show up at the end of the day at his daughter's school red-faced and panting, because he ran there as fast as he could. The practice draws some strange looks. People may wonder what he's running from. But he feels great.

There remains a stigma to conducting intense exercise in public. A good hard sprint does leave you looking a little disheveled. Red face, hair in disarray, panting—it's all liable to attract some looks. Like Phoebe on *Friends*, these people have overcome the self-consciousness that comes with conducting intense workouts in public. They're exercising for the joy of it, and they don't care how they look. But most other people are more like Rachel—they're apt to feel a bit embarrassed to really give themselves over to an all-out workout.

Imagine if it were socially acceptable to bust out a set of burpees while waiting in line at the bank. What if, rather than staring at tablet computers or phone screens while on the subway platform, it became normal to run up and down the access stairs? Or to perform a few pull-ups using the straps on a bus in between a few sets of air squats or push-ups?

Listen, I get it. We're a long way from that. But it does seem like we'd all be a lot better off if we liberated exercise from the fitness center or the workout studio. If we allowed ourselves to conduct exercise virtually anywhere and anytime we wanted. That would be a future to see.

A company in Britain is trying to make that future a reality with a product called the High-Octane Ride, which amounts to a software-equipped exercise bike and a privacy screen shaped like a giant lampshade. The whole thing occupies about the same footprint as a work desk, and it's designed to provide the British public with a sprint workout virtually anywhere—whether the office, the department store, or the park.

The High-Octane Ride uses a form of sprint interval training that employs twenty-second sprints in a format not unlike the one we use in the One-Minute Workout—repeats of all-out intensity and ultrashort duration set into longer bouts of easy cycling.

What I find fascinating about the High-Octane Ride is the way it bills itself as something that nearly anyone can do, anytime, regardless of what they're wearing or what they've been doing.

The workout, the company says, is intense, but over so quickly that its practitioners don't sweat much—so it's possible, and indeed practical, to conduct the workout in work clothes, at the office, between meetings and conference calls. "Because there's no need to change or take a shower afterwards, [High-Octane Ride] is ideal in the office," says the company's marketing literature. "Morning sessions can accelerate fat burning, . . . a session at the end of the working day is a great mood booster." And then there's the line that I love: "A 45-minute jog is so 2014."

As I've explained through the course of this book, research conducted for decades has taught us that exercise provides a wide array of health benefits. In terms of time efficiency, the most potent way to boost fitness involves flavors of interval training—high-intensity interval training and its all-out relative, sprint interval training. Sometimes we don't feel we can block time for a structured workout. And so many people will not bother to exercise at all. On those days, it would be wonderful if we worked sprints into the spare moments of our regular lives. Something like the High-Octane Ride provides us with an opportunity to work out in a relatively private and otherwise socially acceptable manner.

So imagine a future that sees High-Octane Ride's privacy pods distributed anywhere and everywhere people congregate. Right now it's socially acceptable, when we need a break from work, to nip down to the coffee shop for a high-calorie flavored coffee or baked good. Imagine instead a future where, when feeling fatigued, people slip into a privacy pod to blast through an all-out sprint. We'd leave the break feeling exhilarated, and the practice would leave us all a lot more fit, too. Stress and the risk of maladies like cardiovascular disease and diabetes would decrease. Moods would lift.

To spread the word about High-Octane Ride, the company set up a demo model at Harvey Nichols, a tony department store in the upscale London neighborhood of Knightsbridge. A reporter from the *Telegraph* tried it out in a pair of five-inch Louboutin heels. "By the end of the second intense sprint I was out of breath and felt a hint of jelly legs," said the writer, Toni

Jones. "Within five minutes I felt and looked completely normal and was able to head back to the office without changing or even reapplying my lipstick."

One of the things I love about high-intensity and sprint interval training is the way it makes the benefits of exercise available to all. No longer can we claim that we don't have the time to exercise. The excuse would evaporate even more, to overstretch the metaphor, if stations like those provided by High-Octane Ride were easily available. Perhaps utopia via exercise is wishful thinking, but nonetheless it does seem obvious that the world would be substantially improved if all of us were able to get the physical activity our bodies require to be healthy.

The Problem of Nonresponse

Then there are the people who simply don't respond to exercise. How do we reach them? Nonresponders began to be studied in earnest in the early 1980s at Quebec's Laval University. Now an endowed chair in genetics and nutrition at the Pennington Biomedical Research Center in Louisiana, Claude Bouchard was at the forefront of this work. His HERITAGE Family Study put more than seven hundred adults through a twenty-week endurance-exercise intervention. Having conducted experiments for years on pairs of twins, Bouchard was looking to establish that trainability—the ability to improve fitness based on exercise—was a genetic trait; that is, it was something that was possible to inherit. The HERITAGE study succeeded in establishing the genetic nature of trainability. It's something we take for granted today—that someone whose mother or father was a great runner, would tend to be a great runner themselves.

But Bouchard's research revealed something else. He noticed enormous variability in his subjects' training responses. For example, the average improvement in cardiorespiratory fitness was around 25 percent. But a few sedentary people who performed the twenty-week training intervention saw their VO_{2max} *double*. It skyrocketed by 100 percent. Others were classified as nonresponders. That is, they were sedentary people who began and persisted at an exercise program for twenty weeks and their fitness didn't really change much.

Even more remarkable? A small subset of the people in the HERITAGE study, about 2 percent, worked out for twenty weeks and their fitness *actually got worse*. That's right—after five months of supervised endurance training, they actually became *less* fit. Subsequent studies by Bouchard and other researchers revealed the existence of exercise nonresponders for such things as blood pressure, cholesterol, triglycerides, and insulin. For a small number of people, exercise actually *raises* systolic blood pressure, worsens levels of bad cholesterol, and decreases insulin's ability to manage blood sugar levels.

The results of the HERITAGE study, and others just like it, suggested that some people just don't respond to exercise. Now, very few people are *complete* nonresponders. Those who show no change in cardiorespiratory fitness may not be the same people who fail to manifest a change in insulin sensitivity, for example. The greater point is that human beings are tremendously variable. Some people just don't like exercise. And some of those people may be nonresponders—they may not like exercise because they don't respond to it. It behooves the exercise nuts among us to remember that some people have very good reasons for avoiding physical activity.

However, the studies that established the existence of nonresponders all involved *traditional* exercise: continuous, steady-state training conducted for sustained periods, typically between thirty minutes and an hour at a time.

So would exercise nonresponders turn up in high-intensity interval-training studies?

To really attack this question, exercise physiologists needed a big study—similar in size to the HERITAGE study. But the HERITAGE study was an enormous undertaking: an intervention on 742 people spread out over four clinical centers and followed over the course of five months. A comparable HIIT study would cost many millions of dollars today.

In the absence of a HERITAGE-like interval training study, the Mayo Clinic's Michael Joyner and a few colleagues led by Andrew Bacon did the next best thing. In 2013 they published something called a meta-analysis—they did a study on other scientific studies by effectively combining the results. They identified thirty-seven different studies dating from 1965 to 2012 and encompassing 334 subjects. They discovered what you, having read to this point, might have expected. Despite shorter workouts and much shorter time spent conducting hard exercise, the people who performed intense interval training tended to improve their cardiorespiratory fitness more than people who conducted vastly larger durations of endurance exercise. What's more, Joyner and his coauthors concluded that the toughest training regimens, which included interval-training days featuring three- to five-minute sprints *as well as* intense-effort continuous training days, featured "a marked training response" in all subjects—that is, everyone who participated in these protocols improved their VO_{2max}.

Joyner's research suggests that perhaps our previous thinking is wrong. Perhaps it's not that some people fail to respond to exercise. Perhaps it's just that we haven't prescribed them with the right *type* of exercise. What's happening now is a sea change in exercise physiology that's raising the possibility of personalized exercise prescriptions—catering exactly the right type of training to a given person in such a manner that the training summons the maximum possible effect. The idea is to make exercise more efficient to all—and it's coming in the future.

Personalized Exercise Prescriptions

Personalization has already arrived in the fight against cancer, which has become remarkably sophisticated. We've progressed a long way from when oncologists used a small number of different poisonous cocktails to target a large number of different types of cancer. Today one of the first things an oncologist does with a new patient is get a sample of the patient's tumor tissue. The oncologist will look at the tumor's biology, analyze its genome, review various other biomarkers, and then make a conclusion—*this* is the best prescription to fight *your particular tumor.* Some of these approaches use the patient's own immune cells, suggesting there could be as many different approaches to cancer as there are people on the earth.

It's doubtful that prescribing exercise will ever get *that* personal. There are only so many different types of exercise. But physiologists are hoping one day to substantially improve the personalization of workout prescriptions. Perhaps one day soon you'll meet your trainer and maybe the first thing she'll do is take a saliva sample or cheek swab. She'll run it through a portable analyzer and come up with your metabolic profile, which in turn will become matched to the most appropriate blend of resistance, endurance, and high-intensity interval training to get you fit in the most-efficient manner possible.

The idea is to eliminate the idea of exercise nonresponders. Because we've determined exactly the right type of exercise for you—as well as everyone else.

The Exercise Factor

Possibly bringing us closer to a future of personally prescribed exercise is something researchers have sought for decades: a substance that helps trigger the beneficial effects of exercise on the human body. The substance's existence was thought to be the answer to a question that had puzzled scientists for years. Physical activity triggered changes all over the human body, and the changes in turn prompted health benefits that were similarly distributed through many of the body's disparate systems. What triggered these various changes? How could the action of contracting muscles spur such complex and diverse metabolic and physiological benefits? According to a review by Bente Klarlund Pedersen of the Copenhagen Muscle Research Centre, a physiologist named Erling Asmussen speculated that a particular compound might regulate things, which raised the possibility of synthesizing that compound in a manner fit for human consumption: an exercise pill. At a symposium held in Dallas in January 1966, Asmussen called the signal the "work stimulus," or the "work factor." Others referred to it as the "exercise factor."

At first, it was thought that the trigger's source must be endocrine tissue, which would have meant the substance was a hormone. The traditional understanding of hormones saw them as a substance released from an endocrine gland—your pancreas or pituitary, for example—which then circulated through the bloodstream and exerted an effect on some distant target tissue. Melatonin is emitted by the brain's pineal gland to signal to the body that it's time to go to, and remain, asleep. Epinephrine is released by the adrenal glands that sit above the kidneys, to kick the body into a state of alertness, among other metabolic effects.

Beginning in the mid-'90s physiologists began to realize that a substance called interleukin-6 (IL-6) increased as a result of exercise. Not only that—the more intense the exercise, the higher the amounts of IL-6 in the blood.

But what was the origin of the IL-6? What was releasing the stuff? At first, researchers wondered whether immune cells were the source. But analysis demonstrated that the IL-6 that came from certain key immune cells didn't increase following exercise. So was it the liver? Similar experiments on liver tissue revealed IL-6 didn't increase following exercise there, either.

It wasn't until the late '90s that researchers at the Copenhagen Muscle Research Centre grasped something that brought them closer to solving this mystery. By taking biopsy samples of subjects' muscles at rest as well as after exercise, they learned that resting muscle contained very little IL-6. In exercised muscle, though, the level of IL-6 was increased dramatically. Researchers also conducted experiments on rats. They stimulated contractions in one of the rat's hind legs while the other leg rested. Subsequent analysis revealed elevated levels of IL-6 mRNA in the contracting muscle, while levels stayed static in the resting muscle. For final verification, researchers placed catheters in the artery that carried blood to the muscle as well as the veins that transported blood from the muscle back to the heart. And as they suspected, the IL-6 increased upon exercise in the vein that transported blood away from the muscle. The implication? That the IL-6 was being formed in the muscle itself.

Later, eager to learn about the function of IL-6, a Swedish team led by Ville Wallenius altered the genetic profile of experimental mice to *prevent* them from making IL-6. The mice turned out to be obese—and injecting the mice with replacement IL-6 caused the mice to actually lose weight. Researchers had found a candidate for the much-sought exercise factor.

This is really cool, cutting-edge science. In just ten years, our understanding of the way the body responds to exercise has changed profoundly. The same goes for the way we understand how muscles function. Before, we considered muscle tissue to have what was essentially a mechanical function. In average people, skeletal muscle is responsible for about 40 percent of the body's total weight, making it the body's largest organ. And we considered all that muscle to be, essentially, mute. Unable to communicate, existing only to push or pull, and functioning only at the bidding of the activating nerve cells.

We had similar ideas about adipose tissue, also known as fat cells, which were previously considered to be little more than storage containers. Energy sources that were essentially dormant, like a body's version of a coal deposit in the Adirondack Mountains.

Now, our conception of both muscle and fat cells has transformed. Rather than being mute, we understand that both muscle and fat cells communicate in a veritable symphony's worth of different notes. Researchers affiliated with the Copenhagen Muscle Research Centre, who provided substantial contributions to the understanding of IL-6's origins and effects on the human body, proposed that substances released by muscle tissue in response to exercise be known as myokines. Since IL-6, many other myokines have been identified. They have abbreviations like LIF, IL-4, and BDNF, and each one is thought to summon a different exercise response in the body. Muscle cells now are thought to have the capacity to secrete hundreds of different compounds, each of which has its own peculiar communication function. A similar range of complexity exists for the substances secreted by fat cells.

Researchers Bente Pedersen and Mark Febbraio refer to a "yin-yang balance" that exists between myokines and the signals sent out by fat cells, which are called adipokines. When we fail to get the activity that we require, adipokines circulate through the body, encouraging it to hold on to its fat cells—possibly, an evolutionary response designed to help the body cope in times of famine. Adipokines can prevent the body from managing blood sugar correctly. They also encourage the accumulation of plaque on the inside of arterial walls.

Battling adipokines when we exercise are the myokines released by contracting muscle, which trigger all sorts of beneficial health processes. IL-6, for example, is thought to be released when the muscle's stores of glycogen are low. It speeds up the body's ability to burn fat cells and helps the body control its blood sugar levels. IL-15 may help to regulate belly fat. Insulin growth factor 1 and fibroblast growth factor 2 may help build strong bones. Follistatin-related protein 1 promotes the health of blood vessel walls. Still other myokines defend against the development of cancerous tumors. And many of these myokines promote strong muscles from within.

With their talk of yin-yang balance, the health-promoting effects of myokines, and the mortality-boosting effects of adipokines, Pedersen and Febbraio provide clues into the mechanism by which exercise promotes health in the body, and sedentary living promotes its opposite. It is difficult not to see the signals, the myokines and adipokines, as coded transmissions distributed by warring factions in an eternal struggle that happens in the body from birth to death.

Exercise in a Pill?

My longtime colleague at McMaster, Mark Tarnopolsky, has conducted some fascinating research on the dramatic effect that exercise can have on all tissues of the body. One of his remarkable studies, published in 2011, illustrates the way exercise could reverse the normal effects of aging. The study used mice that had been genetically engineered so that they weren't able to repair their mitochondria, the powerhouses of the cells.

As we get older, we grow worse at producing new mitochondria. Say a muscle cell makes a bad mitochondrion. Ideally, the cell would be able to repair the organelle, but as we age, our capacity to conduct such repairs decreases. Tarnopolsky's mice were engineered to be unable to repair their mitrochondria from the very beginning of their lives. Consequently, his mice aged faster than normal mice. Their fur grew gray and patchy much faster than normal. Their sex organs shriveled up. Their skin went wrinkly, and they moved with less energy than normal.

One group of the genetically modified mice *didn't* age prematurely, however, despite the fact that they were bred to have faulty mitochondrial-repair capacities—and that was the mice that exercised. In Tarnopolsky's lab these rodents exercised on one of those little exercise wheels for forty-five minutes a day, three times a week. Rather than being frail, these mice had normal amounts of muscle. Their brains were normal, their hearts normally sized, their gonads were healthy, and even their fur looked lustrous.

The exercising mice also had a lot more mitochondria than they should have, given their breeding. So what did the exercise do? Tarnopolsky has coined the term "exerkine" to describe a factor released in response to exercise that mediates the physical activity's systemic benefits. It's a more general term than "myokine," which refers specifically to compounds released from muscle. Perhaps one exerkine helped keep the mice's fur lustrous and shiny, while another ensured their virility persisted to its previous extent. Or perhaps what did it was the same substance—one single, super-powerful exerkine.

In either case, Tarnopolsky believes that one day soon, within the decade, the exerkines responsible for triggering the age-defying benefits of exercise will be pharmaceutically available, as drugs that are injected into the human body. Tarnopolsky is a world-renowned expert in diseases relating to the cell's powerhouse, the mitochondria. These diseases—with names such as Leigh's syndrome and chronic progressive external ophthalmoplegia—lead to terrible symptoms, such as the reduced ability to move, paralysis, and early death. Many of those suffering from these diseases benefit from exercise, because physical activity promotes mitochondrial health. As a clinician-scientist, Tarnopolsky hopes that exerkine discoveries made in his lab can translate into effective treatments for people with mitochondrial disorders and other neuromuscular diseases.

He also foresees other uses. Therapeutic exerkines could approach the science-fiction concept we mentioned at the start of this book: an exercise pill. Something that provides the benefits of exercise to the people who need it—to enhance oxygen delivery throughout our body, reduce the incidence of cardiovascular disease and diabetes, and stave off the effects of aging.

"The concept of taking a pill to obtain the benefits of exercise without actually expending any energy has mass appeal for a large majority of sedentary individuals," wrote the scientists John Hawley and John Holloszy in a 2009 review on the state of exercise pill research. "It is equally attractive for big pharmaceutical companies that view a potentially huge market and profit numbers."

In reality, it will be difficult to devise a single pill that produces all the effects of exercise. The effects of physical activity are just too numerous and far-reaching. At present, the closest thing we have are chemicals that mimic a specific compound in the body. For example, the chemical known by the acronym AICAR (which I mentioned at the beginning of the book and which, spelled out, is 5-aminoimidazole-4-carboxamide-1-beta-D-ribofuranoside), "mimics" adenosine monophosophate, or AMP, which can trigger some specific exercise-like responses. A month's worth of AICAR injections in rats was shown to boost muscle mitochondria. Similar experiments in mice resulted in a 44 percent increase in running endurance. Some believe AICAR could be therapeutic for treating age-related muscle wasting. The problem, Hawley and Holloszy believe, comes with the long-term use of AICAR, which could actually contribute to muscle wasting.

I share many physiologists' ambivalence to the possibility of an exercise pill, which could bring us closer to a reality where some people believe they don't have to exercise at all. "No matter how far we get with exerkines, or anything else, the best benefits will always come from getting some actual exercise," says Tarnopolsky. "There are so many things that happen when you exercise, it'll be nearly impossible to get *all* the benefits. The pounding and stress that helps bone strength. The higher body temperatures that promote metabolism. The pulling on the tendons. You'll never get any of that with exerkines, or any other drug. That just comes with actually doing the exercise."

Our existence already seems so lazy when compared with that of our ancestors. "A sedentary life is now so prevalent that it has become common to refer to exercise as having 'healthy benefits,' even though the exercise-trained state is the biologically normal condition," Hawley and Holloszy point out. "It is a lack of exercise that is abnormal and carries health risks." Could an exercise pill minimize the health risks that accrue from even greater extremes of indolence? I suspect not, but time will tell.

The Future Is Now

Increasingly, we are realizing the value of physical activity. Now, with interval training, we have a more efficient way to get the required dose of physical activity. Protocols like the One-Minute Workout provide us with a way to boost our cardiovascular fitness faster than ever before. Body-weight interval training provides a mechanism to improve strength at the same time. These days, we can get in a good workout in ten minutes—or maybe even less time.

Today, I use high-intensity techniques for most workouts that I do. That's not to say I don't enjoy a long walk with the dog in the trails near my house. I still do that, but when it comes to exercising for fitness, it's all about interval training. But I recognize that not all people are going to feel the way I do. It's best to consider interval training, and its various less- and more-intense forms, as just another tool to keep you healthy. It's better, in terms of being more efficient at boosting your cardiovascular fitness, but that's not to say that continuous steady-intensity exercise doesn't have its benefits. Plenty of people like to insert their headphones, put on some good music or a diverting podcast, and head off on a run, swim, or cycle—and such calming activities likely provide therapeutic benefits of their own. Likewise, traditional resistance training with weights improves strength. The nice thing is that interval techniques can be incorporated into both.

At the very least, interval training can be used to counteract the vicious spiral that sometimes prevents people from engaging in the exercise that they should be doing. It happens all the time. A stressful day prompts you to put off your workout, which in turn robs you of the anxiety-busting benefits of exercise, which further ups your stress. Which makes you less likely to exercise the *next* day. Soon, several days have gone by. You haven't exercised in a week. You feel fat and out of shape, and your stress is higher than ever.

I felt that sort of stress spiral coming on as I was working on finishing the first draft of this book. It was the holidays. We had houseguests coming, I had work to do, my son had a hockey game to get to, and on top of all that it was my turn to make dinner. Before interval training came along, I never would have been able to get in a workout. After all, I didn't have a free hour to devote to myself.

But thanks to the science of interval training, I knew that individual sprints were able to provide large boosts to my fitness. So at little gaps in the action—after I'd put dinner in the oven, for example, and before we sat down at the table to eat—I simply snuck down to my basement to blast through a sprint session on my exercise bike. The protocol I ended up doing didn't correspond exactly to any predefined workout that I, or anyone else, had studied. But intervals are so variable they can liberate you from the monotony of adhering to one single protocol. I ended up getting in five repeats of hard sprints that lasted sixty seconds each—for a total of ten minutes start to finish, including a warm-up and cool-down. A solid workout despite its brevity, and one that likely served to maintain my cardiovascular fitness similar to fifty minutes of moderate-intensity continuous-exercise would.

Here's the message I hope will resonate with you once you set down this book: Exercise is one of the most important things you can do to prolong your life and improve the quality of your time here on Earth. Interval training can liberate you from the time-intensive demands of exercise. It can also liberate you from feeling you need to conduct your workout in a single block of time.

You do not have to fit your life around your workout. Now, you can fit your workout around your life—sprinkling in sprints here and there throughout the course of your day when you have an opportunity or simply need a stress-reducing break. And remember, you don't have to be a gym rat to use interval-training techniques. You just have to have the presence of mind to increase the intensity of the exercise and then back off—and repeat. Now get out there and get to it.

Acknowledgments

ANY PROJECT OF THIS SORT REQUIRES A TEAM EFFORT and this book reflects the dedicated inputs of a tremendous number of individuals.

Chris Shulgan planted the seed and encouraged me to pursue this initiative. I cannot imagine a better writing partner on what was a true collaboration.

I appreciate the expert help and guidance from everyone at Penguin Random House and The McDermid Agency, in particular Caroline Sutton and Chris Bucci.

McMaster University is the only professional home I have known as a faculty member. I am obliged to the many amazing people who work there.

Studies from my laboratory cited in this book reflect the collective talent, creativity, and effort of a remarkable group of research trainees and collaborators.

I am particularly indebted to Dr. Mark Tarnopolsky for facilitating my research program at McMaster, and recognize the lead roles played by former students Kirsten Burgomaster, Jon Little, and Jenna Gillen. The valued contributions of many others are reflected in the author lists of our published articles.

Much of my research has been sponsored by public agencies; I am grateful for the funding and hope the investment is deemed worthwhile.

Science affords an amazing journey, but it pales in comparison to life as a husband and father. Thank you, Lisa, Connor, and Matthew for your unconditional love and support.

This book is dedicated to my amazing mother, Hazel, and my late father, Larry, who would have been very proud.

Index